Praise for

The Intelligent Gardener

A new book from Steve Solomon is reason for excitement in itself. With *The Intelligent Gardener,* he has re-thought one of the most basic aspects of gardening — basic soil chemistry — and generously supplied us with sensible, practical methods to increase the nutrient density of the food we grow. This book forces serious growers to reconsider some fundamental gardening principles, and to question much of the accepted wisdom on the subject. It's hard to imagine this book not having a significant and lasting impact on the way organic farmers and gardeners grow their crops.

— Mark McDonald, West Coast Seeds

The true test of any gardening book is whether it inspires the grower to new action. *The Intelligent Gardener* indeed inspires me to action, to test my soils more thoroughly, to re-examine my assumptions on compost management and to seek to improve the nutritional value of our produce with a better understanding of our farm's soil. Steve Solomon draws on his years of experience and research to challenge our assumptions of what is good organic soil management and to share his insights for growing the highest quality, nutrient dense food. We are what we eat, and our food is only as healthful as the soil we grow it in. *The Intelligent Gardener* is a valuable tool for anyone seeking to get the highest food value from their garden.

— Darrell Frey, author of *Bioshelter Market Garden*

The Intelligent Gardener is more than just "intelligent," it is bold, it is courageous, and it challenges many of our preconceptions about food, about soils, about farming, and about health. The storytelling is excellent, the science based on experience rather than some out of context lab experiment, the advice and application easy to follow. Everyone should read this, not just gardeners, as it reminds us of where we came from, where we need to go, and provides some clear direction for getting there.

— Michael Ableman, farmer, author of *Fields of Plenty, On Good Land,* and *From The Good Earth*

Gardeners in temperate climates should be very grateful to Steve Solomon for addressing the issues of soil testing for fertility in such an engaging and clear way. I look forward to spending time working with the technical methodology.

— Binda Colebrook, Horticulturist and author of *Winter Gardening in the Maritime Northwest*

the *intelligent* gardener

GROWING **NUTRIENT-DENSE** FOOD

Steve Solomon
with Erica Reinheimer

Cover design by Diane McIntosh.
(clockwise from top left) Chinese lettuce (c) iStock (Mordolff);
Beets (c) iStock (MarcusPhoto1); Spinach (c) iStock (coloroftime);
Carrots, (c) iStock (mrod); Artichoke, (c) iStock (assalve).

Printed in Canada. Ninth printing October 2022.

Paperback ISBN: 978-0-86571-718-3
eISBN: 978-1-55092-513-5

Inquiries regarding requests to reprint all or part of *The Intelligent Gardener* should be addressed to New Society Publishers at the address below.

To order directly from the publishers, please call toll-free (North America) 1-800-567-6772, or order online at www.newsociety.com

Any other inquiries can be directed by mail to:
New Society Publishers
P.O. Box 189, Gabriola Island, BC V0R 1X0, Canada
(250) 247-9737

New Society Publishers' mission is to publish books that contribute in fundamental ways to building an ecologically sustainable and just society, and to do so with the least possible impact on the environment, in a manner that models this vision. We are committed to doing this not just through education, but through action. The interior pages of our bound books are printed on Forest Stewardship Council®-registered acid-free paper that is **100% post-consumer recycled** (100% old growth forest-free), processed chlorine free, and printed with vegetable-based, low-VOC inks, with covers produced using FSC®-registered stock. New Society also works to reduce its carbon footprint, and purchases carbon offsets based on an annual audit to ensure a carbon neutral footprint. For further information, or to browse our full list of books and purchase securely, visit our website at: **www.newsociety.com**

LIBRARY AND ARCHIVES CANADA CATALOGUING IN PUBLICATION

Solomon, Steve

The intelligent gardener : growing nutrient-dense food / Steve Solomon ; with Erica Reinheimer.

Includes index.
ISBN 978-0-86571-718-3

1. Crops and soils. 2. Soils and nutrition. 3. Crops--Nutrition. 4. Soil fertility. 5. Soil mineralogy. I. Reinheimer, Erica II. Title.

S596.7.S65 2012 631.4 C2012-906109-3

Contents

Books for Wiser Living
recommended by *Mother Earth News*

Today, more than ever before, our society is seeking ways to live more conscientiously. To help bring you the very best inspiration and information about greener, more sustainable lifestyles, *Mother Earth News* is recommending select New Society Publishers books to its readers. For more than 30 years, *Mother Earth* has been North America's "Original Guide to Living Wisely," creating books and magazines for people with a passion for self-reliance and a desire to live in harmony with nature. Across the countryside and in our cities, New Society Publishers and *Mother Earth* are leading the way to a wiser, more sustainable world. For more information, please visit MotherEarthNews.com

Acknowledgments

During the ten months it took to write this book, Erica Reinheimer was strongly behind the project. We had endless email exchanges about what might be the ideal target for manganese. She artfully restrained my reckless enthusiasms without triggering any resistance, prevented me from making a few major errors, and contributed a sense of caution that can only come from much experience as a neighborhood soil analyst. Erica read the proofs along with me, and then, re-read them. I can't thank Erica enough.

Erica and I were eagerly helped by the "in-group" posting at the soilandhealth forum; these people repeatedly provided insights and information. Especial thanks to Mike Kraidy, for sharing knowledge from his many years advising farmers. Thanks to John Weil, Lloyd Charles, Norm Cooper, Steve Diver, Jim Karnofski, John Slack, Edmond Brown, Hugh Lovel, Frank Eagan and Roger Martyn. Thank you Gary Kline of Black Lake Organics, for helping me improve my own Complete Organic Fertilizer. Thanks to the many gardeners who took me up on my offer of a free soil test analysis; without their help, I could not have come to appreciate how broad the diversity of soils can be. Finally, this book owes a huge debt to Justin Crawford, D.O. Not long before I began this project, Justin accepted responsibility for (and legal ownership of) The Soil and Health Library. The fact that I no longer had to answer its mail and handle its daily ongoing business liberated enough free attention to let me write *The Intelligent Gardener.* And we all owe Justin a big thank you for continuing the library; in that collection are most of the titles referred to in this book's Bibliography; all available for free download.

Kitchen gardens come and go with the prosperity of the times. During times of emergency, when vegetables are scarce, they become a necessity, and everyone who has access to a small piece of land should feel under obligation to plant a garden. The farm garden is particularly important, because it is possible to produce so much food, for so little effort and with no additional fertilizer. During times of low prices, when farmers' cash is scarce, the properly planned farm garden can supply 75 per cent of the food energy which he would ordinarily buy. This, with what he gets from animals and poultry, should carry him through any emergency.

Subsistence farming is a system of vegetable crop production so fitted in with the production of poultry and animal products, including sheep and wool for clothing, that the family can grow everything it needs on the land, without selling anything. This owner then depends on day labor for sufficient cash income to satisfy his extra needs. It is a system whereby a family can make a living instead of going onto relief when wages are low and work is scarce. It helps a person to maintain his self-respect even though he may be out of work.

— Victor Tiedjens, *Vegetable Crop Production, ca. 1942*

Preface

Erica Reinheimer

I am a neighborhood soil analyst. I help gardeners grow more nutritious fruits and vegetables. I collect soil samples, send them to a lab, analyze the result, then supply and mix the organically approved minerals the soil needs. It's a simple, straightforward small business. Because the result is very effective and my services are not expensive, my customer list grows steadily.

Gardening has been my calling. I like the listening part. I like sitting close to the plants in the morning, being with them, connecting. Understanding how minerals contribute to the garden has deepened that connection. I can't see or smell the minerals, but after soil testing, I know what elements are there, and what has been done to balance the minerals. The intuitive, listening part of me operates even better now.

Like other neighborhood soil analysts, I started helping others after seeing the results of soil mineralization in my own garden. The food from the garden suddenly became even more tasty than it already had been; the garden was far more productive too. Even my green beans (yes, green beans!) became flavor-packed delights. My own produce is now so much more flavorsome that I am disappointed in the taste of the organically grown veggies at our local farmer's market. I even grew great-tasting tomatoes in a cabbage summer. "Geez, this is so easy, compared to the other efforts I have put into the garden," I thought. "Every gardener should be doing this."

I began gardening with compost 30 years ago, and I still use it. I still advise starting a new garden by applying compost. Good compost improves tilth, adds air to the soil, holds water and fosters life. Great stuff. In those early years, I realized something was lacking, but I didn't know what else the soil needed, if anything, and I didn't know what was in the compost. So I added more compost. Is it any wonder it usually takes years and years to get a garden really going?

Later, I began using soil tests. This worked better, but I didn't fully trust the recommendations I was getting from the lab. I used the results as a guideline, and tried to guess at what organic materials my garden needed. Then I discovered the art of soil mineralization. Now, for the first time, I have an effective *system* for converting land into garden. Now, all plant nutrients are abundant and available in the right proportions. Now I have really superb results. Mineral balancing provides a great foundation for compost to work and for the soil biology to thrive. Foundations are a great place to start.

Putting minerals into soil guided by a soil test is no guarantee that the plants will be able to access them. But, it is certain that if the minerals are not in the soil (or in foliar sprays), the minerals will not be in the plants. All the other parts of gardening are just as important. There is a famous aphorism which says, "Growth is controlled not by the total amount of resources available, but by the scarcest resource." Please, let the scarcest resource be something like the gardener's time, not some bit of mineral lacking in the soil.

I am fortunate to be in the vanguard; I expect that balanced remineralization will soon be commonplace in backyard gardening. It works so well that if you'll just give it a try, you will wonder why you didn't do this before. Your plants will thank you, and your family will thank you, and if you become a soil analyst yourself, your neighbors will thank you.

Skepticism is a healthy attitude when it comes to taking garden advice. As a relative newbie in 1977 I was avidly reading back issues of *Organic Gardening* magazines going back to the time of J.I. Rodale. I confess I developed a certain amount of arrogance because I used the Organic Method. About the same year, Steve Solomon was homesteading in Oregon. He was eating mostly from his own garden, growing his food according to the organic principles I was reading about. But, he did not become supremely healthy. At times, he says, he felt like an old man (he was in his late 30s). On this diet, the condition of his teeth worsened greatly, and he began losing them.

Of necessity at that time, Steve's diet was mostly his own organically grown vegetables. But his soil was seriously lacking certain minerals and was hugely overdosed on others. Pretty standard — we see imbalances in every soil we work with. Consequently, the taste and nutrient-density of Steve's vegetables were far from what they could be, and needed to be, for him to be healthy. Most home gardens are not all they could be for the same reason, but most home gardeners do not make the majority of their diet come from one piece of imbalanced soil. So they never find this out. Steve will tell you how he came to gradually discover how to effectively mineralize soil and how he thereby restored his health.

Steve Solomon is the right person to be teaching this art. His late wife, Isabelle Moser, practiced nutritional medicine, and by their association and common interests, they connected the health of our soils to the health of our bodies. Steve is the chief architect of the Soil and Health online library, one of the great repositories of soil and nutrition on the planet. He has the unique talent of being able to notice the obvious while the rest of us are just taking things for granted. The connection between the place we live, the climate and the soil, the food we grow and our health seems obvious, but only after it has been pointed out.

I first encountered Steve through a mail-order business he started, Territorial Seed Company, then through his book, *Growing Vegetables West of the Cascades*. In that book, Steve concentrated on the unique advantages and challenges of gardening in the Pacific Northwest. By focusing on one distinct climatic region, he not only made the book more relevant for gardeners in Cascadia, but firmly planted the notion that all gardening is local, and that all gardening advice should be too. It is up to us readers to see how an author's ideas relate to our own locations.

So much of good gardening is about appreciating the unique potential of your garden's climate and soil. As you turn the pages of this book, you will find insight and advice that will further your understanding of your own situation. This is why this book is important — so that you may see your own garden with increased depth and clarity.

This book is for gardeners and homesteaders; it uses organic methods. But it widens the scope of organic gardening to include some of the best techniques used by today's certified organic farmers. The result is better, more productive gardens. More nutritious food. And, the best tasting vegetables you have ever eaten.

Erica Reinheimer
2012

Introduction

North America wallows. Doctors, dentists, hospitals, clinics — rake in money. The costs of an ever-growing fraction of the population requiring care and/or supervision are sinking the economy. Australia follows blindly down the same path, a few steps behind. But I'm not talking about money here; it's unnecessary suffering that concerns me. Maybe more than that — it's the ever-lowering quality of daily life.

We've known the remedy since the 1930s but have not yet applied it. The simple fix — grow nutrient-dense food — is not mentioned, while powerful interests profit from the current farming system. Their conscience-less apologists, the most credentialed scientists money can buy, spew convincing statistics and cite carefully front-loaded peer-reviewed studies. Consequently, the public mind is enormously confused about what constitutes a healthy diet, about differences in the nutritional value of foods, as well as about the worth (or lack of it) of the attractive-looking but nutrient-poor foods being offered them.

I assume you already are concerned about what food does (or doesn't do) to make you healthy. You are seeking better answers, you are seeking better food, and almost for sure, you are uncertain.

But perhaps you are not confused. Suppose you fully realize nutrient-dense food cannot be bought reliably at any price and have set out to grow enough yourself to make a meaningful difference. You have no desires contrary to that aim. No conflicts about it. And what you want from my book is to just get on with the details of how to do it. You may have already created a large garden or are about to, or are about to expand a few small beds into something significant to the family economy. And you want to do it right. For you, I suggest (after reading this Introduction) skipping forward to Chapter 5, where you will learn how to take a soil sample, how to analyze the soil audit you'll get back, and how to work out a soil prescription that will offer your vegetables balanced nutrition. You can then proceed to grow nutrient-dense food. In Chapter 5, you'll get an arithmetic-powered fertility target generator that'll tell you exactly what each unique soil needs.

The rest of you are uncertain. You wonder if this remineralization stuff really is the way to go. Maybe you've read a few issues of *Acres* or *Mother Earth News* or subscribed to Rodale's *Organic Gardening* for a year or three, as so many of us oldies did when first starting out. Consequently, you are now aware of conflicting viewpoints...and most of them sound pretty reasonable. Some prominent voices assert the following pleasing fiction: there is no need to pay attention to soil minerals because almost every soil already contains all the elements needed to grow nutrient-dense food crops; currently non-productive soils only need the right sort of biological encouragement — the sort that can inexpensively be brewed up out of some specially concocted compost (but only if you buy a patented brewer and the appropriate starter cultures). Or maybe you have been enthusiastically told by a devotee that Fukuokasan had the right idea. Or you have been inspired by the organic gardening ideal that asserts there is nothing like compost: compost is the remedy, compost is fertilizer, compost is life itself. Another half a dozen gardening systems come readily to mind. New gardeners running that gauntlet often achieve inner peace by selecting a guru-system whose notions resonate with their basic predispositions or existing opinions, and thenceforward, cling to that approach like a lifesaver in a stormy sea of confusion.

I propose to help you put these inevitable confusions into perspective and help you settle on an effective plan for producing a nutritious abundance. And, just in case you are concerned that this old reprobate might lead you from the politically correct path, here's my bottom line: when a soil is very far away from offering plants an abundant and balanced supply of minerals, if key plant nutrients are nearly missing or way out of proportion, then the food it produces cannot be nutrient-dense. Period, full stop!

There is also a biological side to it, equally important to plant health and ultimate food quality. So which is the chicken and which the egg? I say soil minerals come first. If first you bring the minerals into proper balance, then the whole soil ecology, all the microlife — the worms, nematodes, algae, amoeba, fungi, bacteria, both helpful and harmful — all those living things come into a healthy balance too. In my opinion, when it comes to microlife, there is rarely any need to import them. When the soil favors the proper organisms, they *will* predominate, appearing as if from nowhere. As Louis Pasteur admitted on his deathbed, the body's inner chemical nature is everything, the bacteria is nothing — disease organisms appear because the body has become a welcome home for them. Same with soil.

Microorganisms that naturally dominate in balanced soil work to effectively release plant nutrition that had previously been locked up and unavailable. They also assist the crop to assimilate that nutrition. Soil microorganisms can manufacture enough nitrate nitrogen to make a garden independent of nitrate imports. Biology can enormously forward an already balanced garden soil; biology can perform miracles. But biology will only do its job with extreme effectiveness after you have fed the soil to satiation and brought it into balance.

The first chapters of this book aim at helping you to re-evaluate food-growing information you may have already acquired. Suppose you arrive at Chapter 5 a few evening's of reading behind those who are already convinced that remineralization is the way to go. Once you're a bit softened up by my passionate prose, what are you in for?

First, you will be asked to get an inexpensive soil test. The highest-priced American lab I recommend charges $20. The cheapest lab on my list charges $14 (as of 2012). For small fees like these, a soil audit will provide no personalized recommendations. You'll promptly get back a computer-generated form reporting the amounts of plant nutrients present. With this book's help, you'll be able to work out a list of soil amendments targeted at bringing your soil into balance. It's a matter of applied arithmetic, made simple.

First you'll learn which mineral elements your soil needs and which of the many possible OMRI-listed (Organic Materials Review Institute) materials to use to make that happen effectively. Then I am going to explain how to make effective compost. I've been making compost for 40 years now; for the first 35, I didn't get the kind of terrific results the old-timey organic gardening books led me to believe I would inevitably achieve. Maybe I can help you skip over those 35 years of practice.

So what are you getting into if you choose to remineralize? In short: several years of rapidly improving results until you achieve a high organic matter level and the best mineral balance your soil is capable of. The specifics of garden remineralization vary with location and circumstance. The goal is pretty much the same, but the route varies. Chemically, soils can be extremely different. Almost all of them are out of balance, at least to some degree. The majority of soils seriously lack essential plant nutrients. And it is not unusual for soils to have nutrient excesses, sometimes really big ones. How much time, material (and money) it may take to restore missing plant nutrients or to reduce excesses, varies. Some soils cannot hold on to large quantities of plant nutrients, so they can be transformed rapidly by the application of small doses. This facility to be easily transformed also allows soils to untransform equally rapidly, which is a major obstacle when it comes to growing nutrient-dense food. Other soils (heavy soils in particular) can soak up a great deal of plant nutrition, so it can take a larger quantity of soil amendments applied over several years to get to the levels you want.

But you will not have to wait several years to see results. Not at all. Unless you already have a magnificent garden and trying to upgrade it is like gilding the lily, remineralization will bring immediate, major, massive improvements. Even if you are a new gardener or are starting a brand new garden, you will start having excellent results the first spring — maybe terrific results, if your soil did not start off too far out of balance.

Soil minerals are strong medicine, even garden stalwarts like dolomite lime are powerful amendments. And there are few things more important than a food garden. So can you trust the advice of someone who does not possess an advanced degree in horticulture or a license to prescribe?

About the Author

I have spent 35 years making erratic progress at fending off disease through dietary reform; so far, I've managed to keep two steps ahead of the Piper. I also treasure personal independence. After four years of serious suburban backyard food growing, I decamped to an Oregon homestead. I've lived that lifestyle for the past 35 years. Homesteading suits me well. I spend as much of my time and energy as possible pursuing my own interests and aspirations. I can't help but protest when forced to pay attention to things that do not interest me — in other words, I never successfully worked for someone else without soon becoming terminally bored. In a similar way, I find formal education irritating — a waste of my time that could have been better spent on self-directed study.

I did manage to achieve a right and proper BsEd (in history) from a certified state university. But it took me seven turbulent years of starting, quitting in disgust, and then restarting six months or a year later; the cycle repeated several times until I realized the process was never going to be enjoyable or personally meaningful, so I just got on with it to get the degree as quickly as possible. I have no formal agricultural training. I had no formal business education either, yet in the 1970s I managed to bootstrap (with next to no starting capital), build up, and then sell a thriving book production business. No one exactly taught me how; I just picked it up through my dealings with tradespeople and customers.

High school inorganic chemistry was one schooling experience I *do* value. If I had not learned — honestly learned — inorganic chemistry in the 11th grade, 55 years ago, I don't think I could have written this book. Happily, because I did learn it, I can explain the subject at hand in a way that won't require you to know chemistry or even high school math.

When I went into the mail-order seed business, I found myself dealing with qualified agronomists and plant breeders who were amused to hear me spouting half-baked ideas acquired from Rodale's *Organic Gardening and Farming* magazine. Their bemusement prompted me to do some serious study of horticulture and agronomy on my own. I think I did pretty well for an amateur. I do have some agronomic lacks:

I do not know organic chemistry, so the intricacies of plant physiology at a chemical level and some of the more complex soil chemical reactions are beyond my comprehension.

On the other hand, I have 40 years of hands-on, serious food gardening experience. (The modifier "serious" means that I attempted to make vegetables a majority of my diet.) And I have been teaching others to garden better since 1979, when I wrote the (admittedly primitive) first edition of *Growing Vegetables West of the Cascades*. If you want to do some serious gardening, I can help you. The art of remineralizing soil to increase nutrient-density was developed by independent biological farm advisors working in the tradition of William Albrecht, a pioneering researcher in the relationship between soil fertility and human health. Farm advisors can acquire lifetimes of experience in a few short decades by analyzing other people's soil and seeing the results of their advice. The Bibliography lists a few books written by people who were highly successful at this. I have had the opportunity to chat with a few of these guys, which is amazing in itself, because boy, are they are busy! They travel widely. And they hang around with big farmers who spend (and hope to make) big bucks.

Fortunately there is at least one advisor, Michael Astera, who does focus on small growers. Michael wrote a book, *The Ideal Soil*, that gives amateurs a simple way to analyze their garden's soil without having years of experience or a degree in horticulture. It is a mathematical system that adjusts the proportional relationships that should actually exist among plant nutrients. The method allows the amateur to know — about as well as any practicing biological farm advisor knows — how much of each plant nutrient should ideally be in their soil. These "ideal" plant-nutrient targets are compared to a soil test report that shows the amounts that actually *are available* to plants. The difference between what is available and what is ideal is made up for by the addition of fertilizers. (Or, the comparison lets you know which nutrients are present in excess.) Easie peasie. The approach has one other powerful attraction: when you achieve "the ideal soil," you should also achieve the highest possible nutrient-density in the foods you're raising.

A few years ago, Michael began participating in "soilandhealth," an Internet discussion group I moderate. I had never before thought that such precise soil balancing needed to be applied to the home garden. On the forum, I had complained of "tight" compacted soil despite the addition of lots of organic matter. Michael suggested I change the type of lime I was using. I did, and a year later my soil was loose. At his suggestion, I got a soil test; his analysis helped me to get results beyond any expectations. So I read his book. And then I closely studied his book. And I kept on studying. I was inspired — at age 69, no less! I started giving free soil test analyses and fertilizer prescriptions to anyone on the soilandhealth forum who asked for them. I started doing them for people in my neighborhood. Before long, I had become a local garden soil analyst — with a half-dozen large bags of assorted fertilizers in the garage.

Participants on the soilandhealth forum generally refer to an arithmetical system like the one I explain in this book as the "Astera Method." Constitutionally, I can't be a true believer in anyone's system; as I studied his, I began to introduce my own tweaks. To his great credit, Michael himself isn't a true believer in his own method. He freely states that his targets and the proportionate relationships that generate those targets are educated, inspired guesses. But contemplation of what might constitute the ideal soil involves a big playing field — one with enormous variability. So, based on my 40 years of hands-on-hoe experience, I have come to disagree with Michael in some respects. That's not unusual. Every successful farm advisor out there has a slightly different opinion about what constitutes a perfect soil prescription.

When people apply the art of balancing minerals to an existing garden, they are often inspired in the same way Erica Reinheimer was inspired — in the same way I was inspired. Some of my readers will soon want to help their whole neighborhood. And thus it is that I foresee the birth of a new helping profession — the neighborhood soil analyst. It's a microbusiness requiring investment in little more than a dozen or so farm-sized bags of plant nutrients and an accurate scale. A soil analyst assesses a garden, orchard or field, takes soil samples and sends them in for analysis, works out a soil prescription, supplies the fertilizers (if only a small garden is involved), and provides consultation as needed. All for a modest fee. If a few hundred people start doing this because of reading my book, I'll be proud of myself for having written it. If a few thousand start, it will be a major social transformation because tens of thousands of people will discover for themselves that health really does come from better nutrition.

Good health and good gardening to you and yours!

Steve Solomon
Tasmania, 2012

Chapter 1

Why Nutrient-Dense Food?

Would you be skeptical if I told you people could normally live past age 100, die with all their original teeth, be in sprightly enjoyment of life up to their final weeks, and all this could happen if only we fertilize all our food crops differently? Skeptical? Most people think I am harmlessly mad.

How to achieve and maintain health is a scary, important topic. People get upset when health opinions they believe to be facts are challenged. I've not found it easy to change health-related or disease-curing opinions. Not even with statistics. I've got plenty of numbers supporting the case that eating nutrient-dense food produces long life and good health — even extraordinarily long life and unusual good health — but scientists-for-hire can always out-statistic an amateur, and people these days have been made so insensitive to facts and figures, it is pointless using numbers as a tool to convince. When I contemplate that long chain of utterly brilliant people who, since the 1930s, have all failed to convince the world that health equals nutrition divided by calories, well, if they all failed, what chance have I? I am no scientist. I am not a lawyer. If convincing is needed, I think the very best thing for me to do is to relate my own experiences and observations.

I was instantly hooked by my first vegetable garden. The activity itself was calming and centering (these days I'd use the term "balancing"); it made me smile. Gardening still makes my heart sing. I can't give you a credible explanation for why it does that. But for an incredible one, I'd say it's a relaxation technique for karmic warriors on long service leave.

After 40 years of serious food growing on five different soils in two quite different North American climates and on two soils on a remote South Pacific island so unique it's almost a nation unto itself, one thing prominently stands out: my average physical condition went up and down according to the soil I was eating from. The most prominent (and worst) period was nine years of eating mostly my own organically grown vegetables produced on an infertile Oregon Coast Range soil. This period probably cost me my teeth, although I did not lose them all right away.

1973, age 31. My first food garden was entirely and unreasonably over the top. It occupied the rear half of a one-acre house block in the western part of southern California's San Fernando Valley. This valley is typical of semi-arid regions; it's a fertile flat of fresh young soils that recently (geologically speaking) washed off the surrounding mountains. The West Valley seemed a near-perfect place to live in the early 1970s. The air was still free of smog. The soil grew things well, and the neighbors did not jump my fence to pinch produce. What more could I want from suburbia? In short order, vegetables became a major part of what we ate; homegrown vegetables largely replaced meat-and-potatoes. I became a confident food gardener, bored with restaurants, and I dreamt about escaping the Los Angeles rat-race.

I have grown a substantial food garden ever since. I can't imagine living differently. And knowing what I know now about the nutritional qualities of supermarket stuff, if I want to stay healthy, I have little choice but to make my own vegetables most of what I eat. I invented a word to describe my lifestyle: *vegetableatarianism*. The word does not mean that animal foods are excluded. A vegetableatarian is someone who's trying to repair the damage caused by harmful food addictions by eating mostly vegetables, cooked and raw.

Prior to vegetableatarianism, I had been visiting a dentist every 12 to 18 months to have a few new cavities filled and my teeth scraped clean of thick, rock-hard deposits; I had already had two root canals and a bridge. A few years after home-garden vegetables became a major part of our total food intake, I noticed that I had developed no new cavities in a good while. The chemistry of my mouth had become inhospitable to decay organisms. Unfortunately tooth *decay* was not my only dental situation; I had lost a lot of jawbone.

During my first 30 years, and especially during my childhood, I was malnourished. I'd been bottle fed (on doctor's advice, infant formulas were at that time considered scientific — far superior to breast milk). That was not a good start. I recall eating Velveeta (cheap syntha-cheese) melted on macaroni (devitalized semolina) and Velveeta cheese and mayonnaise sandwiches on white bread with one iceberg lettuce leaf included. Or else peanut butter and jam on white bread. These were typical take-to-school lunches. Cream of Wheat (devitalized semolina) cereal for breakfast, lots of pasteurized homogenized milk (which I was allergic to and suffered greatly from drinking, but no one, including my drug-pushing pediatrician, made the connection), meat-and-a-starch dinners, and not much in the way of vegetables. There was rarely anything that these days I would consider a proper salad. Instead, I was offered a bit of watery iceberg lettuce and a thick slice of tasteless tomato with prepared mayonnaise dressing thick atop. Is it any wonder I didn't like tomatoes? I remember my mother did bake tinned green beans in a casserole with creamy onion cheesey stuff and crispy bits on top. Oh, and there were snacks, lots of snacks, especially when watching TV after school; in the evening, there were salty snacks, like potato chips dipped in mayonnaise, and sweet snacks, like ice cream or cookies dipped in pasteurized homogenized milk.

Consequently, my growing body received inadequate minerals and, in its wisdom, to prevent loss of my ability for fight-or-flight (which requires big skeletal bones), my body had to short-change non-essential bones. As a result, my face was long, narrow, and angular, even though my genetics called for it to be broad and flat; some of my teeth came in crooked because my jaw (a non-essential bone) never grew large enough. My body had been forced to steal construction materials (calcium and phosphorus) from its own jawbone to use for other essential purposes.

1978, age 36. My wife and I left Los Angeles to homestead in the Oregon Coast Ranges. Having eaten for some years from well-mineralized soil that was not too far out of balance, I was in quite "good nick:"

Fit. Energetic. Happy. Full of interest. Our main goal was to be free and clear after getting set up. Our dream property was 20 to 40 acres, about half of it sunny, cleared land, and half a healthy woodlot. And, of course, I dreamed the homesteader's dream of having a strong-flowing, year-round spring that would nurture it all. As it turned out, we could only afford five acres of worn-out hillside that had once grown winter wheat. Decades of autumn plowing followed by heavy winter rains had resulted in loss of *all* the topsoil. Descendants of those responsible for that crime still lived across the road, fattening calves on unerodable bottomland that was fast losing its drainage. Note that they fattened, but weren't breeding their herd — the reason being that cattle failed to reproduce successfully on their exhausted fields. On my side of the road there remained two feet of infertile, acidic silty-clay subsoil down to bedrock on the gently sloping parts; even less soil was left on the steeper parts of my hillside.

I was confident that if we completely avoided debt I could create some kind of part-time small business that would pay the taxes, keep us driving an old beater, and let us clothe ourselves. And that was my main intention: a self-sufficient lifestyle in which I did not have to sacrifice my best hours and energies to making money. Besides, I was supremely, stupidly confident I could quickly convert any old clay pit or gravel heap into a veritable Garden of Eatin' by putting in plenty of organic matter and lime. *Organic Gardening* magazine had repeatedly asserted I could do that, and the several dozens of veggie growing books I had closely studied — many of them published by Rodale Press — reinforced the idea.

The garden fence enclosed the largest area I could defend against deer. I used two 100-yard-long rolls of field fencing topped with two strands of barbed wire — thereby enclosing a square that was 74 feet on each side and 7 feet tall. The first winter, we ate a lot of dried vegetables and cooked beans, which was eating like we were homesteading in Ohio, not Oregon. But then, most of my neighbors ate that way, too. In the second year, I learned how to grow cold-tolerant greens all winter and how to hold mature root crops in the ground from autumn until spring. The garden was supplying about half our calories for about eight months a year and about a third of them for the remaining four months. I was certain that as I gained more skill at winter growing, I'd increase that percentage.

In late 1979, I went into the mail-order vegetable seed business. To do that business ethically and responsibly, I had to conduct variety trials to evaluate suitability for a family kitchen, organic gardening (in rather poor soil) and Oregon's climate. The trials garden was one-half acre — this was in addition to the family garden. In the beginning, I did not intend to eat much from the trials. I certainly did not have the time to harvest, pack and sell the surplus; the trials were grown for information only. I intended to toss the cucumbers and zucchini into the paths to rot or let my employees take home however much of the surplus they wanted.

But that is not quite how it worked out. Financing the business required investing every remaining cent of my savings. For my first two years as The Seedman, I could not support myself in the modest style to which I had already become accustomed. So I stopped spending money, ate even more from the trials garden and did not feel hard-done-by for having this chance to improve my health.

After a few years eating mostly my own vegetables, I found I was losing energy. And my teeth were worsening. The teeth did not decay; I started having what my dentist called "wobblers," loose teeth that eventually fell out by themselves if they didn't get too painfully infected first. My body again had to rob non-essential bones of the calcium and phosphorus it wasn't getting from the vegetables I was eating. By 1983, the seed business could support us at about the official poverty line. By its fourth year, 1984, Territorial Seed Company had become nicely profitable; we felt economically secure — at least it was security as we thought of it in our early 40s. But managing a fast-growing business was getting tiresome; I'd been making major efforts for six straight years and could now afford to relax a bit. So in mid-June, 1984, immediately after the trials garden had been established for the summer, I took myself, my new wife Isabelle and her 12-year-old daughter to the English-speaking tropical South Pacific island of Viti Levu (Fiji) for a sabbatical. There, we rented an inexpensive furnished apartment in the capital city of Suva. Isabelle's daughter went to the International School while we hung out, intending to relax for up to six months while I polished up the third incarnation of *Growing Vegetables West of the Cascades*.

In Suva, we ate more or less our usual vegetable-and-fruit diet with some delightful substitutions, such as a local form of raw-fish ceviche and local papayas and mangos. After a few months, our health began to improve. My wobbly teeth tightened up by themselves. Isabelle's fingernails got hard again. We recovered our energy and enjoyed an ongoing sense of well-being. Why, we asked? Was it lack of stress? The climate? The food? Life was certainly less stressful living in a tropical climate during its cool season, but I couldn't say for sure if that was the source of our better health. But when I did a little investigation into the food we were eating, I discovered that almost all the produce in the Suva public market came from one place, the Sigatoka River Valley, less than a 90-minute drive from Suva. So I rented a car for the day and booked a visit at its agricultural experiment station.

Fiji has a two-season climate much like the Big Island of Hawaii. Sigatoka farmers raise temperate-climate vegetable crops during the cool season. In May, temperatures moderate, the rains stop, and the gentle, constant trade winds resume. For the next six months, Fiji's climate is like summer in Oregon, but with balmy nights. Vegetables grow excellently. In November, days turn hot, humid and less windy. By December, the trade winds stop completely; the stagnant air feels heavy. Sweat drips from your body even when you are sitting quietly in the shade dressed as the locals are, with only a bit of thin cotton cloth wrapped around your middle. We became lethargic. By December, the temperate vegetable crops in the Sigatoka had all died of heat- and moisture-induced diseases; our diet became much less interesting. Next came the cyclone season. Even if there are no major cyclones (what Americans call hurricanes) in a particular year, there are still many heavy thunderstorms; most of the year's rain falls from December through April. This can be the "starving time" in traditional Fijian life, especially so for several months after a cyclone strips gardens bare.

Even if there is no cyclone, the crops still die of heat and humidity, and the fields are taken over by weeds and rank grasses. If it does prove to be a year when a cyclone comes, the river floods the entire valley, depositing silt and sand. Once the rains stop, the new soil particles and the chest-high growth of grasses and weeds are plowed in. These weeds serve as the major source of soil organic matter, and the freshly ground rock deposited from the floods restocks the soil's mineral nutrients. A research bloke working there asserted that farmers in the Sigatoka Valley use no fertilizer; frequent additions of silt suffice. The vegetables coming from that valley are sprayed because no matter how well-nourished the plant, temperate-climate species cannot handle some tropical insects. But fertilized? Never. Given compost or animal manure? Never. It was on these poison-sprayed, unfertilized never-given-compost-or-manure vegetables that my wobbly teeth tightened and our health swiftly improved. This contradicted everything I thought I knew about growing good food. So, as soon as we got back to Suva, I dove into geological surveys and discovered that the watershed of the Sigatoka River was mainly ultrabasic igneous rock. Eureka! And thank Serendipity for anticipating this moment, having put me through an inspiring university-level geology class.

Igneous rocks come from liquid magma — volcanism. Geologists classify igneous rock into three general types: acidic, basic and ultrabasic. The distinction has to do with the mineral composition of the magma that made them. Acidic igneous rocks are usually light in color, contain large quantities of silicon (quartz), potassium and sodium, but not much else, and they are less dense (lighter weight) than basic or ultrabasic rocks. The best known sort of acidic igneous rock is granite. Soil forming out of acidic igneous rock has an acid pH; it is not particularly rich in plant-growth nutrients. When I think of the effect of eating from granitic soils, what comes to mind is the narrow, pinched faces of upper New England.

Basic igneous rocks are darker in color and weigh more. They contain less silicon (quartz), less potassium and less sodium than acidic rocks, but hold large amounts of calcium and magnesium and higher overall levels of plant-nutrient elements like phosphorus and sulfur. Basic igneous rocks usually develop into effective agricultural soils that — in humid climates — are only mildly acidic. The best known basic igneous rock is basalt. The biggest exposure of this sort of rock I have experienced personally is the Old Cascades. The roots of this ancient chain of volcanoes can still be seen in a few spots in western Oregon; the flows from these volcanoes cover most of eastern Oregon and Washington.

Ultrabasic igneous rocks are rare. They are quite dark in color, heavy and dense; they are rich in metallic plant nutrients like iron, manganese and copper, as well as carrying a lot of calcium and magnesium. The richest upland agricultural soils derive from this sort of parent material. The richest alluvial soils are those that derive almost exclusively from ultrabasic igneous rocks. Because there are no extensive regions covered by ultrabasic rocks, no large river systems carry a load of only ultrabasic silt, but the Sigatoka River does. The Sigatoka Valley probably has better soils than Egypt had before the Aswan Dam was built. It may have the best soils this side of Hunzaland.

The Universal Force handed me a mystery — we had experienced a health resurgence while eating mostly poison-sprayed vegetables grown on soil that never saw amendments of animal manure or compost, much less any chemical fertilizer. And on that food, my wobbly teeth tightened, Isabelle's fingernails got hard, our hair grew faster, our energy improved. Our attitudes improved. And then came December. It got uncomfortably, exhaustingly, depressingly hot and humid. The third edition of *Growing Vegetables* had been completed, and the seed business was on the telephone demanding my attention. So, we returned to Oregon and resumed eating from my organically grown garden and trials ground. After less than one year in Oregon, my teeth were again loosening, Isabelle's fingernails were again softening, our overall attitude and energy level was again declining. And I had been given a huge gift — I had discovered there was something important for me to learn because what happened to our health in Fiji contradicted my faith in organic gardening as I understood it then.

The Organic Religion

The serious gardener strives for better results next year. We study constantly — learning by observation, by experimentation, and, in our formative years, from garden magazine articles and a motley assortment of books, usually with the term "organic" on the cover. Most organic gardening books convey the same basic principles, and the new gardener comes to think likewise. For example, organically grown food is always said to be more nutritious than conventionally grown food. Yet, after six months in Fiji, we were obviously better nourished and a lot healthier than we had been eating mostly organically grown vegetables.

I enjoy giving garden lectures. After Fiji, I made public confessions about my wobbly teeth and the results of eating from my organically grown trials ground. My confession must have made it safe for other homesteaders to come forward — privately, in confidence — to share that they, too, had lost many teeth or otherwise had a significant lowering of their overall health after eating primarily from their own organic garden for some years. Our mutual disappointments were not the consequence of our food having been grown organically. They were the consequence of the food having been raised in soil that was not minerally balanced.

After reading the above, I'll bet many of you just switched off. You are so sure that developing dental problems from eating organically grown food is impossible that you dismiss my assertion. And that's why, before I explain how to produce the most nutritious possible food *using organically certifiable techniques,* I first must attempt to disabuse you of certain commonly held notions about organically grown food. And then I must profoundly impress on you that fact that mineral balancing is merely a natural extension of organics, not a disagreement with it.

Pro, con, or indifferent, you hold opinions about organically grown food, about organic farming, and about organic gardening. I request that you take a moment to step back from those opinions and have a good look at them. If your opinions favor organics, you should know that the information that shaped them almost certainly originated from J.I. Rodale's *Organic Gardening* magazine and the many books Rodale Press has published. Rodale's vision was so powerfully and confidently presented that Rodale's Organic Doctrines are now accepted by most contemporary garden writers. Oldies like me learned their basics straight from Rodale; the following generation of garden writers learned their stuff from my cadre and from Rodale; Rodale still carries on.

Rodale firmly established a set of positive feelings and opinions about Organic. Product brand names were similarly promoted. I'd bet Warren Buffett one of the world's wealthiest financial manipulators and investors, would pay an arm and a leg to have purchased the organic franchise way back when. Starting in 1942, Rodale Press magazines and books repeatedly asserted, suggested, inferred and implied both subtly and overtly, both straight out and between the lines, that organically grown food is far more nutritious than chemically grown food — a half-truth. J.I. Rodale was an ideologue at heart. Absolutely certain about the rightness of his own opinions. If J.I. didn't agree with you, your name was never mentioned in Rodale publications, and the gardening public never discovered you. That is why most gardeners these days have never heard of William Albrecht.

J.I. Rodale's agricultural opinions powerfully impressed the North American psyche. *Organic Gardening and Farming* magazine had a circulation of 1.4 million in 1980, which is when I began renting their subscriber lists for my seed business. In that same era, Rodale Press's main profit earner, *Prevention* magazine, had several million subscribers.

The soil-fertility building methods our current batch of food gardening books recommend are little changed from Rodale's Press's 1940s doctrines. These methods have been stated and restated — and repackaged with little alteration by garden writers ever since. This repetition has continued so long and been so widespread that the ideas have come to be considered almost scientific truth, the very ground we can confidently stand upon, because Everybody Else has said it so many times before.

Today's organic gardener needs to catch up. Contemporary certified organic farming uses far better agricultural science than what was available in the 1940s. That's because organic farmers and market gardeners are disciplined by being in business. They must make a profit, and to do that they must be efficient producers of good-looking food. By the 1990s, some organic vegetable farmers had built big businesses, and money buys political clout to influence the rules defining acceptable organic practice.

Ironically, as organics morphed into an industry able to charge higher prices (and make higher profits) because it owned J.I.'s mental franchise, Rodale's dogmatic belief system was found to be ineffective. Efficient practices that J.I. would have condemned for ideological reasons are now allowed by organic-certification bureaucrats because they actually *are* in harmony with healthy biological agriculture. An example: certified organic producers are now allowed to use a limited range of chemical fertilizers that do not harm soil life or the soil itself. But most organic gardeners still believe *all* chemical fertilizers to be artificial substances and so, "of the devil."

Rodale's original organic gardening system was built on the following articles of faith:

- Organically grown food provides far superior nutrition compared to conventionally grown stuff; it produces great health and well-being in those who eat it.
- A successful organic gardener builds soil fertility mainly by importing organic matter, and, to a far lesser extent, through the importation of natural rock flours — especially of lime. Nothing should go into the soil that has been chemically processed or otherwise altered from a natural condition other than being finely ground up.
- Almost all soils are capable of growing super-nutritious food in abundance. If the soil is not performing, the reason is that it lacks organic matter. The presence of more organic matter increases the rate that nutrients, previously locked up and unavailable to plants, are naturally released. Get the soil biology sufficiently active (by adding compost and/or manure), and it will release enough nutrients to grow good crops.
- You can't possibly have too much organic matter. Organic matter provides needed plant nutrients. Organic matter lightens up the soil, loosens it, "builds it up," as the old timers say; the fluffier the soil remains during the whole crop cycle, the better the plants grow. Since earthworms eat organic matter, soil fertility is best gauged by the earthworm-per-shovelful method. So compost and/or manure should be repeatedly spread several inches thick.
- In humid temperate regions, the soils are naturally acidic, so lime is used to bring the soil pH close to neutral. Soil pH is the only essential test needed; liming is done according to this test result.

And if lime is to be spread, the best sort is dolomite because dolomitic lime contains both calcium and magnesium. (Later, I will show you how excess magnesium brought in via dolomite causes loads of problems.)

- Chemical fertilizers are unsustainable, their manufacture and transport needlessly wastes energy resources, their use inevitably damages soil microlife and kills earthworms. They also deplete the soil of organic matter. Use them, and soon you'll have to spray chemical poisons on your sick plants.
- The way to distinguish positive, good, useful soil amendments from harmful, negative, evil ones, is by their naturalness. If the substance occurs naturally, it may be used to build soil fertility. If it is a highly mineralized rock, it may be ground to a fine powder so it more rapidly decomposes to feed the soil, but no chemical processing of these rock minerals is acceptable. Anything that comes directly from the soil can be used as an organic fertilizer, including animal manure and crop waste. Organic materials are allowed to be considerably more processed than rock-based minerals; they may be composted or even chemically processed and still qualify for use. (In this last respect, I am thinking of oilseedmeal, which is what is left over after the oil is squeezed or more usually, chemically extracted from oily seeds.) Processed (ground, dried, heated) slaughterhouse wastes, like bonemeal, bloodmeal and meat meal are highly desirable soil amendments. (This is one article of faith that did get reconsidered — once fear mongers raised the issue of Mad Cow disease.)

With a bit of sly amusement I point out here, that under those rules, sodium nitrate and potassium chloride, both being naturally mined, soluble fertilizer salts found extensively in Chile, and borax, which is mined in Death Valley, California, are suitable for use in organic gardening. But sodium build-up from using sodium nitrate can be quite dangerous to soil; and there is good evidence that chloride fertilizers resulting in the rapid leaching of subsoil calcium — ruining soil for a long, long time. Borax is still accepted, but I do not think these others are allowed any longer. On the other hand, calcium nitrate, an entirely synthetic fertilizer, is a wonderful substance when circumstances call for it, but certified organic growers are not allowed to use it. Same with monoammonium phosphate.

The Rodale Organic Doctrine is easy to comprehend. Everybody can use it confidently. It proposes that if you make and spread plenty of compost, and (impulsively) select soil amendments from a list of approved substances, and avoid those not on the list, you're being socially responsible, can take pride in being good to the environment, and your food will turn out to be highly nutritious. But in truth, it is possible to organically grow food that is as devoid of nutritional content as the conventional, industrial stuff. Home gardeners do this all the time — at least it's fresher, that's something. The worst of it is that most organic gardeners believe themselves to be on the side of the angels and the environment and their vegetables are the greatest thing ever. As the saying goes: It's not what you do not know; it's what you think you know that isn't so.

I have tried to fairly represent the organic belief system, albeit with a bit of my bemusement showing through the cracks. However, take my word for it: if you grow your own food that way while paying inadequate attention to your soil's mineral balance, the probability that your garden's vegetables will be supremely nourishing is supremely slim.

Nutrient-Dense Food

Achieving a nutrient-dense diet involves perfecting three things. First: some entire food classes are more nutrient dense than others; we need to avoid foods with little intrinsic nutritional content. Second: some batches or lots of the same kind of food can be far more nutrient dense than others. These differences can be due to genetics, but usually have more to do with the soil on which the foods were grown and sometimes at what stage of maturity they were harvested. Finally, some foods have been devitalized, that is, processed so as to reduce their nutrient content. White flour and refined vegetable oils are two glaring examples.

Different productions of the same type of crop can vary greatly in nutritional quality. The same variety of wheat can have very different protein levels depending on the soil and, to a lesser degree, according to the amount of rainfall that year. Some varieties of same kind of vegetable have far higher levels of vitamins and minerals. So, the same is not really the same.

Another class of differences in nutrient-density is between types of food. This conversation is sometimes termed "making healthy choices." For example, wheat usually is far more nutrient dense than rice. In fact, rice is probably the least nutritious of the major cereals, especially white rice. So if it were possible to choose between rice and some other grain, it might be wise to avoid rice.

I distinguish between nutrition and fuel. We benefit from almost unlimited quantities of nutrition, but excesses of fuel burden the body and become deposits of fat. Both fatty foods and sugary ones are highly concentrated forms of energy that carry little or nothing in the way of minerals, vitamins or enzymes. Even raw honey, the best natural sweetener, has barely enough minerals and enzymes in it to justify its consumption; for sure, cane sugar does not. It contains nothing but energy.

Practicing healthy choice also means avoiding devitalized foods. To be healthy, our bodies need every bit of nutrition they can possibly assimilate. If, for convenience or for profit, nutritional content is removed or destroyed during processing, the consumer's health gets shortchanged. Health really does equal nutrition divided by calories; devitalization removes much of the nutrition, but few of the calories. In fact, devitalized foods usually become more calorie-dense as they are made less nutrient-dense. Much has already been written about this situation; it is an example of commonly held knowledge most people choose to ignore.

Making healthy choices extends beyond the simple selection of wheat over rice, or brown rice over white rice, or the avoidance of unnecessary fat and sugar. These days, the choice has to be made based on invisible differences. Most varieties of wheat can, if grown on properly fertile ground, contain quite a bit of protein, many minerals and key vitamins. In order to contain enough gluten to make decent bread, wheat must be 14 % protein or more. However, there's a type of wheat used to make soft, white instant noodles. It contains about 8% protein; it was bred to grow on soils of low fertility and is less nourishing than even white rice. Both sorts of wheat look much the same until you try to use them, but you can't make rubbery bread dough out of noodle wheat. Even otherwise high-protein hard red wheat grown on unsuitable soil might end up at 11% protein. In ideal conditions the same variety might reach 19%. At 11% protein, the stuff is termed "soft wheat," good for little but making crumbly cake or for chicken feed. At over 14% protein, it is termed "hard wheat." At 16%, it becomes highly prized by bread bakers and sells for a premium. At 18%, it's a baker's treasure. Same variety; higher protein; entirely different nature. These kinds of differences occur in all foods.

People hugely underestimate the importance of nutrient-density. I am entirely without a footnote for that assertion, but still, it's obvious. If people really did value nutrient-density, they would not be making the kinds of food choices they routinely do make. The hard, unappealing truth is that the average nutrient-density of your entire food intake over your entire lifetime is the basic cause of your current state of health or disease. The next most significant contributing factor to your current physical state was the nutrient-density of your mother's nutrition from her conception to the point that she stopped breast-feeding you (if she did breast-feed). The main exceptions to this are environmental pollution and poisoning with workplace/agricultural chemicals.

In about 1990, I invented a simple mathematical formula to express the idea just described:

$$\text{HEALTH} = \text{NUTRITION} \div \text{CALORIES}$$

I did not invent the concept my equation expresses; that universal law was proved beyond all doubt by multi-generational animal-feeding studies during the 1920s and 30s. Unfortunately, this vitally important truth has been conveniently ignored ever since by senior medical authorities controlling institutions such as the AMA and the state licensing boards it controls, the CMA, the Australian Medical Association, etc. Acknowledging that truth wouldn't have been good for business.

The decade of the 1920s was a time of enormous scientific advancement in the fundamentals of biology, health and agriculture. We discovered vitamins, developed the "newer knowledge of nutrition," and learned to measure (assay) some of the nutritional qualities of foods in a laboratory. The existence and nature of vitamin deficiency diseases was first revealed by Dr. Robert McCarrison, who in 1922 published *Studies in Deficiency Disease*. The book was developed from animal-feeding studies done in his own laboratories. In that same decade, Dr. Francis Pottenger did landmark multi-generational cat-feeding studies, with results so meaningful that ordinary people interested in holistic health still talk about them. Dr. Pottenger established a control group of properly fed cats that were entirely free of disease and then, by giving several generations of these cats improper feeding, induced, and then, by several generations of proper feeding, reversed, the same sorts of disease and degenerative conditions commonly found in humans. Often these diseases are incorrectly attributed by medical doctors as the result of unfortunate genes. They aren't — Pottenger's properly fed cats almost never exhibited deformities, but their poorly fed progeny did.

In the 30s, McCarrison observed that populations of lab rats change their size, overall health, longevity and social nature when fed the various diets of India over several generations. Some groups waxed large and healthy and long-lived; others shrank, shriveled, became ill-tempered, and stopped breeding. The studies were reported in two major medical-school lectures McCarrison presented in 1938, one in Pittsburgh, the other in England. Slides were shown, evidence presented, and then the whole topic was swept under the rug. Interestingly, one Pittsburgh attendee was J.I. Rodale. You can read McCarrison's lecture online at the Soil and Health Library.

And in that same era, Weston Price, DDS, took an interest in preventative dentistry. Around 1900, young Dr. Price left his native North Dakota to practice in Cleveland. Although Cleveland was a place of great financial and social opportunity, Price took more interest in prevention research than repairing teeth. However, he couldn't determine how the nutritional connections worked because, as he put it, he lacked a control group. Yes, he would have an occasional patient with excellent teeth. But why did this person have such good fortune? And what, if anything, could those with poor teeth have done to prevent their condition from developing? This puzzle was especially confusing because an extraordinarily healthy and long-lived person sometimes thrived on a diet of overcooked red meat, potatoes stewed in greasy gravy and whiskey. The only way to scientifically work these confusions out is to first establish a healthy control group and then see what happens when something different is applied to part of that healthy control group. The problem was, there were no groups of people in or around Cleveland, or even in or around the entire United States, that had consistently healthy teeth. And if such a group could be located, how could a researcher get them to agree to control their diets? Or trust that they had actually eaten as promised?

Fortunately, in Price's era, people still existed that did possess excellent teeth. They all lived in highly inaccessible places. These folks were to become Price's control groups. Starting around age 60, Dr. Price went traveling with Mrs. Price to see what they might discover. They journeyed to Europe, Africa, the wild north of Canada, the west coast of South America, Africa, Australia, New Zealand, Polynesia and Melanesia (Fiji). Their dental connections opened doors; local health authorities were enlisted to guide the Prices. Guide? Why guide?

Before World War II, remote communities still existed that had no access to the foods of civilization. No village store sold white flour, marmalade, sugar, tinned sardines. None of that. These peoples survived almost entirely on what they hunted, fished for, gathered or grew locally. The visiting Prices conducted mass dental examinations and developed statistics on the incidence of caries (tooth decay). They searched the communities for the sick people and, through interviews, developed an impression of what diseases were routinely faced. The Prices took excellent photographs, most of them showing facial bone structure, and sometimes, wide-open mouths. They drew correct and highly useful conclusions about why these people were so healthy. By 1939, when he published *Nutrition and Physical Degeneration,* Price had learned almost everything needed for us to transform this planet into a healthy place. If only we, collectively, had wanted to do that. If only those with political and economic power had been willing to lead us in that ethical direction.

I have derived one huge and highly liberating principle from Price's book — *there is no ideal diet for homo sapiens.* Or more accurately stated: if there is an ideal diet on which humans can have average life spans exceeding 120 years, then we're a long, long way from discovering what it might be — and it probably has more to do with the soil foods come from than which foods are chosen. Every one of Price's remote communities was entirely healthy and long-lived (as we think of long life these days), but each one depended on different basic foods. In the far North, people mostly ate animals and fish supplemented with berries and other wild vegetables in the short midsummer period when they were available. Some healthy communities were primarily vegetarian, eating garden produce and cereals. Isolated South Pacific islanders — Melanesians and Polynesians both — depended on seafoods supplemented with garden vegetables, semi-wild fruits, and coconuts; the Gaelics, of the often-foggy Outer Hebrides, mainly ate seafoods and oats, with a bit of extra-hardy garden vegetables, like kale. In a remote Swiss valley, Price visited extraordinarily healthy people who depended on rye bread and dairy products.

All these primitive communities had excellent overall health. Except for the heavy meat eaters of the far North, who only experienced a fully enjoyable life into their early 60s, they had life spans equal to or better than Americans or Canadians have now. In all these communities, the people — the old people — possessed all or nearly all their own teeth; Price found extraordinarily little evidence of decay and no gum diseases. Anyone with missing teeth had lost them through trauma. Price did frequently find traces of tooth decay in individuals who had spent a few months away, living on town food. But, when they returned home, their teeth healed themselves; new enamel, somewhat like scar tissue, formed over the pits. Chipped or broken teeth also healed themselves, as the body's Designer intended.

Every healthy community Price visited — which included humans of every color, shape and hair texture, who had many types of lifestyles and ate many types of diets — was found to be composed of good-natured, honest, responsible people possessed of an innate spiritual awareness that did not require regular church attendance to awaken. Their women did not fear childbirth, did not suffer much during it and rarely died from it. And the reason for their health: food with high nutrient-density. In *Nutrition and Physical Degeneration* Price, considering the sad health situation of the United States, wrote:

> *It will therefore be necessary for an adequate nutrition to contain approximately four times the [US Government recommended] minimum requirements of the average adult if all stress periods [like childbirth] are to be passed safely.*
>
> *It is of interest that the diets of the primitive groups which have shown a very high immunity to dental caries and freedom from other degenerative processes have all provided a nutrition containing at least four times these minimum requirements.*
>
> (pp. 274–276)

Michael Astera analyzed Price's data and concluded that in the communities Price visited, the average intake of calcium was 5.1 times the US recommended intake; for magnesium, it was 13.6 times. The average healthy "primitive" was ingesting 5.4 times the amount of phosphorus, 17.4 times the iron, and more than 10 times the amounts of what Price termed "fat soluble activators," which we now know as vitamins A, D and E.

I urge you to purchase a copy of Weston Price's *Nutrition and Physical Degeneration*. Study it. I suggest re-reading it every five years. If you are uncertain about making personal dietary reforms, Price's data will reinforce your wavering will. With Everybody Else irresponsibly eating junk food, someone trying to eat properly needs support, and support is what you'll get from Weston Price. When I lecture, I always suggest that *Nutrition and Physical Degeneration* be placed on the shelf next to the family bible — and that it be consulted as often or oftener. The book's true power comes from its 100+ black and white photographs. A picture truly is worth a thousand words. After you compare the people in Price's photos with people in your neighborhood, in your supermarket, in your family — in your bathroom mirror — you'll be thoroughly convinced that: 1) we have significantly degenerated from what humans are meant to be, and 2) most or all of your neighbors and family members are far from being healthy specimens. You, too, most likely.

I wish I could create the full impact of Price's photographs by reproducing a dozen of them; but, to be a transformative experience, you must see them all and read the captions. Price's photos reveal that natural ethnic differences in surface appearance barely conceal the underlying truth (which is bone structure). All *healthy* humans look much the same beneath their skin — broad, wide, rather flat faces, with wide jaws that have plenty of room in them to hold all the teeth. Because the face is broad, the nose spreads out flat, looking strong and sturdy instead of thin and delicate. Peasant-like. Healthy women usually have a full pelvis. Those narrow-faced, hipless females we consider fashionably attractive these days can barely reproduce. I always come

Fig. 1.1: *A typical "black house" of the Isle of Lewis derives its name from the smoke of the peat burned for heat. The splendid physical development of the native Gaelic fisher folk is characterized by excellent teeth and well formed faces and dental arches.* CREDIT: PRICE-POTTENGER NUTRITION FOUNDATION, WWW.PPNF.ORG.

away from Price's book looking at people differently — noticing their jaws and teeth and the width (what Price termed "development") of their mid-face — instead of their clothing or how they make themselves appear.

Our modern foods, be they what I term "industrial food," or home-garden produce, mostly fall far short of providing enough nutrition to make us truly healthy. If the fundamental foods of industrial agriculture were grown so as to become more than half as nutritious as

Fig. 1.2: *Above: brothers, Isle of Harris. The younger at left uses modern food and has rampant tooth decay. Brother at right uses native food and has excellent teeth. Note narrowed face and arch of younger brother. Below: typical rampant tooth decay, modernized Gaelic. Right: typical excellent teeth of primitive Gaelic.* CREDIT: PRICE-POTTENGER NUTRITION FOUNDATION, WWW.PPNF.ORG.

they could possibly be, most of the diseases currently ruining peoples' lives would vanish by themselves. We could eliminate most livestock disease the same way.

Your descendants could be as healthy as Price's primitives. A nutrient-dense diet that began well before your daughter's conception and continued uninterrupted at least until your grandchild was weaned (assuming both live without much stress or other chemical insults) could extend longevity such that your grandchildren, or, for sure, your great grandchildren would live to age 100+. There is good evidence suggesting it would be 110 years. They would die still possessing all their teeth, would enjoy well-being and have good energy without regularly being medicated for degenerative diseases like high blood pressure, diabetes, circulatory disorders, cancer, etc.

Dr. G.T. Wrench, author of *The Wheel of Health*, had a term for how the mother's state of nutritional health influences the child. He called it "the start." It is vital that a body start out with a full nutritional complement. Wrench, expanding on the animal-feeding studies of McCarrison, explained how it takes a few generations of proper feeding to fully charge the body with nutrients. Women who grow up eating highly nutritious food have bright, intelligent children who rarely develop the so-called inherited diseases or have birth defects. Wrench said if our wheat, milk, meat, fruit and vegetables were grown so that they were as nutritious as we knew how to make them (in 1939), then to achieve an enormous transformation in average health, people would have to do no more than make reasonably healthy food choices most of the time.

I assisted my previous wife, Isabelle Moser, a gifted nature-curist, in healing many diseases — even serious, life-threatening conditions — by teaching people to thoroughly reform their diet. However, the fruits, vegetables and unprocessed whole cereals of today's industrial agriculture (including organic industrial agriculture) have been grown on depleted soils that do not produce nutrient-density. Consequently, healing through dietary reform and detoxification, which had a long track record of working for pre-World War II alternative healers, now often fails to heal serious diseases. I opine that this is why most contemporary natural healing centers use raw food diets. The body actually does extract more nutrition from raw foods with less effort, leading to healing. However, judging from Weston Price's data, if our basic dietary feedstocks were grown so as to be maximally nutrient dense, we could heal diseases (and stay supremely healthy) while eating mostly cooked food.

Organics advocates assert that food grown in compliance with their rules must be highly nutritious, health promoting and far more nutrient dense than conventionally grown foods. They assert that organically grown plants are genuinely healthy plants that rarely (if ever) are attacked by insect or disease. If you need to spray insecticides and/or fungicides your plants lack nutrient-density; when you can grow food without spraying, your harvest is, by definition, nutrient dense.

The certified organic method can achieve high nutrient-density, but usually it doesn't. The "conventional" method rarely produces nutrient-dense food, but it could. And the marketplace offers no incentive for producing maximally nutrient-dense food — not at all. There is a big financial reward for obtaining more bushels or tons per acre while reducing production costs to the absolute minimum. There is a reward for achieving perfect appearance. But the market is only beginning to recognize nutritional quality as something people are willing to pay a premium for. And to make it even more complicated, trying to understand the creation of nutrient-density by applying the organic vs. conventional distinction — it just doesn't work.

Glossary

From this point forward, soil science comes into the conversation. There is a fundamental learning principle that ideas themselves are rarely incomprehensible; they are easy to grasp. But if the words used to express those ideas are not understood, or, worse, misunderstood, then the ideas themselves seem confusing. So, to help you avoid this obstacle, here are concise definitions of the technical words used in this book that are not in common use. There are 21 entries for 28 words. Please read this list through twice, carefully. That should do it.

- *Available/Unavailable:* Soil nutrients are *available* when in a form that can be readily taken in by plants. Some nutrients are dissolved in the soil solution; these are easily taken in by plants as they take in moisture. Cations and anions (defined two entries further on) attached to clay or humus are also available. Plant nutrients present in soil in insoluble forms are *unavailable.*

- *Capillarity:* Moisture in the subsoil has the ability to rise toward the surface through this principle. Plants lift moisture from root to leaf through thin capillary tubes. Capillarity's ultimate lifting limit determines the height of the tallest tree.
- *Cation/Anion:* Anions and cations are atoms possessing a faint electrical charge. Cations have positive charges; anions have negative charges. These charges attract or repel each other much as magnets do, allowing atoms to hook together into chemical compounds. For example, sodium, Na^+, a cation and chlorine, Cl^-, an anion, attract each other, and combine to form NaCl, table salt. When salt dissolves in water, it breaks apart into a sodium cation, Na^+ and a chloride anion, Cl^-.
- *Divalent/Monovalent:* Cations and anions can have more than one faint electrical charge. The *valence* is the number of electrical charges they have. Atoms can also have one, two, three and sometimes four charges (also referred to as having a valence of one, two, three or four). The valence number is indicated by the number of plus or

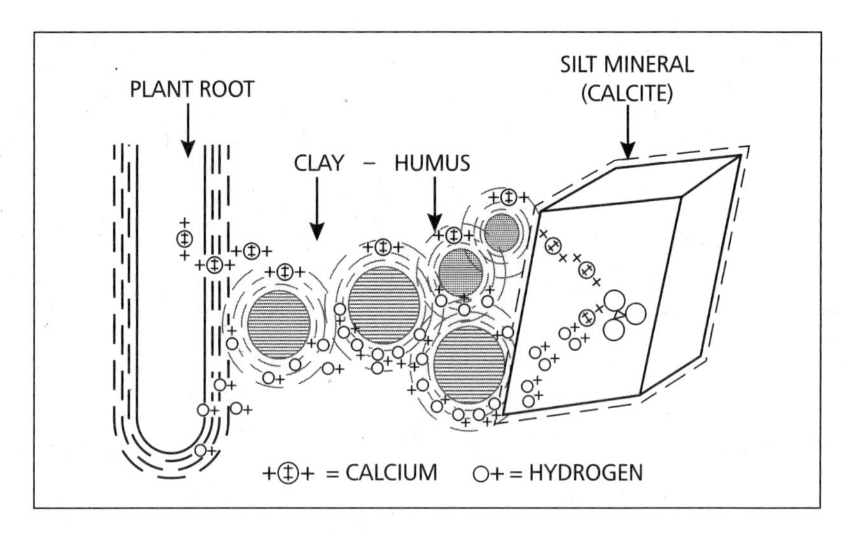

Fig. 1.3: *Cation exchange capacity illustrated.*

minus signs attached to letter symbols of the cation or anion. For example, Na^+ is a monovalent cation; Ca^{++} is a divalent cation. I'll be using the terms anion and cation a great deal; you won't see the term *valence* again in this book, but the underlying concept is useful for you to have.

- *Extraction method:* When testing soil, a precisely weighed sample of finely ground, completely dry soil is soaked in an extractant solution. Depending on which extractant is used, it dissolves some, most or all of the plant nutrients present. Then the extractant is analyzed, the amounts of nutrients in it are determined, and thus the mineral composition of the soil can be estimated.
- *Evapotranspiration:* The combination of evaporation, which is the total amount of moisture lost from bare soil or open water, and transpiration, which is loss of moisture from (mostly) the leaves of plants. It's just a fancy way of saying "all the moisture the soil loses."
- *Flocculation:* Clay can shrink up tight, into an airless, sticky mess that'll grow nothing, or else it can loosen up, act more like soil, develop a crumb structure, let air in, and allow excess rain to pass right through it. When it loosens, it is said to have *flocculated*. Clay is best flocculated by attaching a great many calcium cations to it.
- *Furrowslice acre:* The classic moldboard plow turns over a slice of topsoil 6 to 7 inches thick. This is called a *furrowslice*. One acre of topsoil 6 to 7 inches thick is a *furrowslice acre*. In soil testing, this volume of soil is often assumed to weigh 1,000 tons, or 2,000,000 pounds.
- *Igneous rocks:* Rocks that formed from volcanic activity as opposed to sedimentary rocks, which form on the seabed from deposits of sand, silt and clay coming out of the mouths of rivers, or, in the case of limestone, from chemicals in sea water originally leached from the land. Igneous rocks usually have far higher levels of plant nutrients in them than sedimentary rocks have, except for limestone, which is nearly pure plant nutrient.
- *Jillion:* An imaginary number that is unimaginably larger than one trillion.
- *Leaching:* When a large quantity of water enters the soil (when it rains hard), moisture flows downward through the soil, taking with

it minerals dissolved in the soil solution. These dissolved minerals often get transported beyond the reach of plant roots. Usually, *leached* minerals end up in the groundwater; from there they flow into the ocean, making the sea ever-more salty.

- *Light/Heavy soils:* Bits of mineral, sand and silt do not have permanent electrical charges and cannot hold on to cations. Clay and humus do have permanent electrical charges that can hold on to cations and anions. Soils vary greatly in their capacity to hold cations and anions. Those with a small capacity are *light soils* and those with a larger capacity are *heavy soils*. In this sense, "light/heavy" does not exactly mean the physical density or weight of the soil; it is not that a clay soil is necessarily heavy or that a loamy soil is necessary light, although that's usually the case.
- *Milliequivalents:* Understanding this word is crucial to understanding this book. So read this definition over a few times, and think about it a bit. Make sure you grasp it. The soil's ability to hold on to cations is measured in *milliequivalents* or *meq*. It's just a number. One milliequivalent represents an incredibly large quantity of permanent, charged attachment points in a certain amount of soil. But you do not have the job of estimating how many of those points exist in a furrowslice acre or of counting the number of cations in a bag of fertilizer that will stick on to those attachment points. Your task will just be making use of the number representing that quantity (which will be provided in your soil test report). You will learn how to convert this number into weight of fertilizer. Much of soil analysis consists of matching the amount of fertilizer going in to the capacity of the soil to hold on to it. Suppose a furrowslice acre is only capable of holding one meq; if that entire meq of holding capacity were to be exclusively filled with calcium cations, that furrowslice acre would hold 400 pounds of calcium. More on this later.
- *Mineralized:* When soil is given plant nutrients in the form of fertilizers, organic concentrates, crushed rock flours, etc., we say it is being *mineralized*, or *remineralized*. When soil naturally has a high level of nutrients, we say it is highly mineralized.
- *N-P-K:* These letters are chemical shorthand symbols for the three nutrient elements most commonly put into fertilizers. *N* means

nitrogen; *P* stands for phosphorus; *K* is potassium. (The letter "K" derives from the Latin word for potassium, *kalium.*)

- *pH:* Water consists of one positively charged hydrogen cation (H^+) combined with one anion called "hydroxyl" or $(OH)^-$, so the proper chemical formula for water is HOH, but we usually use H_2O. Water is not a super-stable compound; it easily splits into its component anions and cations. When there are free hydrogen cations in water (H^+), it is *acidic.* The more H^+ there is, the more acidic the water is. The concentration of hydrogen cations is measured as *pH.* When there is no acidity, i.e., no H^+ present, the pH is said to be 7.0, the neutral position on the pH scale. As the pH number declines, the concentration of hydrogen cations *increases.*
- *Saturation percentage:* Soil has a limited, fixed capacity to hold on to cations. Most of the spots available are usually (and should be) occupied by calcium cations. If 80% of the soil's total cation exchange capacity (defined below) is filled by calcium cations, then the calcium saturation is said to be 80%.
- *Soil development:* As soil forms, it goes through a steady process of losing minerals due to leaching; this process is called *soil development.* A well-developed soil has lost much of its original mineral content.
- *Soil solution:* The moisture present in soil contains all sorts of dissolved plant-nutrient cations and anions. This nutrient-laden moisture is the *soil solution.*
- *Sufficient/Deficient:* Sufficiency is a commercial agronomist's concept. If feeding a crop more of a plant nutrient does not result in any increase in yield, then the amount of that nutrient (its level) is deemed *sufficient.* To whatever extent the bulk yield is lessened by the lack of some nutrient, the soil is *deficient.* Sufficiency is all about achieving peak yield without economic waste. This book does not target sufficiency; it targets nutrient-density.
- *TCEC/CEC:* For our purposes, *TEC* (*Total Cation Exchange Capacity*) is the number of milliequivalents a furrowslice acre is capable of holding. Light soils, by definition, hold few meq; heavy soils hold many meq. *CEC* (*Cation Exchange Capacity*) usually refers to how many meq a pure substance, like a specific kind of clay, or a

type of humus, can hold. A typical soil, consisting mainly of sand and silt and a small percentage of clay and a few percent of humus, may have a T(otal) CEC of 15 meq in a furrowslice acre.

Chapter 2

History from a Nutritional Viewpoint

I think history pays far too much attention to politics, wars, kings, generals and leaders — and their crimes, follies and mistakes. Other factors have far more influence. Like agriculture. Civilization itself is possible only when there is a reliable production of large food surpluses, which gives people free time to specialize in trade and manufacturing, to create music and literature, and also to make war far more effectively. Yank that surplus food out from under a civilization, and one thing inevitably happens: it collapses. A civilized region can be conquered and reconquered; its infrastructure can be demolished during conquest, and its population can be decimated, temporarily. But if the agricultural base remains capable of producing great surpluses, the same civilization will quickly reappear, albeit with different tax collectors.

The schoolbook history of western civilization begins in Mesopotamia about 7,000 years ago. It starts with the early Tigris-Euphrates city-states and moves on to Egypt, and then to the various cities and empires arising around the Mediterranean — Crete, Greece, Phoenicia, Rome. Every early civilization had a similar pattern: first came the exploitation of rich, virgin soil in semi-arid or arid climates; this led to vigorous, aggressive, productive, expanding populations that dominated each region. In semi-arid or arid climates, the soil is little leached and often highly mineralized; such land (if it doesn't slope too much) can grow excellent food for a long time after first being put to the plow, maybe for centuries, even without fertilization. Eventually, though, unsustainable population growth causes massive soil depletion and/or the outright destruction of a soil resource, leading to insufficient agricultural productivity to feed a large population. Despite efforts to maintain a system by robbing one's neighbors, a civilization lacking its agricultural base inevitably collapses.

The immediate causes of soil depletion differed from civilization to civilization — slightly. In Mesopotamia, food production depended on irrigating a densely settled floodplain that these days is called Iraq. Farming in that near-desert required a complex network of canals coming off the Tigris and Euphrates. But overpopulation led to denudation of their watershed, leading to massive soil erosion that gave the Tigris and Euphrates a heavy silt load even when it wasn't flooding wildly. The silt steadily filled in the irrigation canals, forcing constant dredging; as the watershed declined, ever-more effort had to be put into cleaning the canals. Despite wars of enslavement waged to get labor to keep the canals open, agriculture became ever-less productive, until it collapsed. Once that happened, it no longer was possible to feed the millions that had lived there, and the region became the sparsely settled semi-desert it still is.

In Italy, rich limestone-derived soils on valley floors and flats produced abundant high-protein wheat that fed an aggressive population, while forests on the mountain slopes maintained springs and rivers during the rainless summer and moderated flooding during the rainy winter. But population pressure pushed people into the uplands; iron tools allowed them to readily clear the forests. So they terraced and planted crops on these slopes or grazed them. Once woodcutters and goats had done their jobs, soil washed off the slopes and was carried away in winter floods, filling in the valleys with silt, making them into swamps where mosquitoes and malaria reigned. Gradually, the Mediterranean region lost its ability to feed a large population and turned into something more resembling desert, even though it still gets about as much winter rainfall as it ever did. Unfortunately, there is insufficient soil remaining on the hills to soak up this moisture, so when it does rain hard, water now runs off in horrific floods; the springs, streams and rivers have become more seasonal than permanent. When the empire's food system broke down, and the much weakened population could no longer organize sufficient resistance, wild Germanic tribes were able to flatten the western part of the Roman Empire.

Egypt was different. Geography prevented surplus population from pushing into the Nile's watershed, so its precious annual flood continued unchanged from ancient times until the Aswan Dam was finished. The Nile's flood reliably carried large quantities of mineralized silt that replenished adjoining fields. The ancient historians said Egypt was so fertile that one grain of wheat yielded 200; perhaps this was a bit of a brag. In these modern times, it is normal to harvest 60–80 bushels per acre from one bushel of seed grain; before we had chemical fertilizers, the North American average yield was in the vicinity of 40 bushels per acre. I fear for Egypt without the Nile's flood. The overall health of the Egyptian people is already in swift decline.

Europe did not rise by exploiting an untapped, hugely fertile soil resource. After the Roman Empire collapsed, western Europe muddled through a long Dark Age. Around 1400, Renaissance aristocracy took up additional interests beyond pillage, war and theft — applying science to make their estates profitable. Roman farming practices resumed, which included liming, the rotation of crops, and the use of long fallows, or leys, which means putting the field to grass for some years while it rebuilds its fertility. Europe started producing nutrient-denser food in larger quantity. This surplus, combined with superior weaponry and the increased human vigor, helped thrust Europe into world dominance.

The New World

When the English began colonizing North America, they found what seemed an agricultural gold mine requiring only clearing the forest of trees (and of Native Americans) to put into production. And was that ever a forest! In 1600, the eastern North America old-growth hardwood forest was not all that different from what's left of the Amazon's. In 1600, it was possible for a squirrel to ascend a tree on the east coast of Virginia and travel all the way to the Mississippi River without once having to come down. Native Americans living east of the Great Plains were gardeners who supplemented their diets by hunting, fishing and gathering. But their population was small, mainly, I think because the

First Nations lacked metals, especially iron. So native gardens were only won out of the bush with great labor; they were few and small. The leached forest soils of eastern North America did not support large herds of grazing animals, so hunting was not a dependable way to support a large population. Scattered wild nut trees didn't produce huge yields. I don't buy "lack of vaccination" or "lack of previous exposure"; if anything can explain how so many natives died in plagues brought by the English, it is their dependence on a diet largely made of starchy foods (corn and winter squash) grown on depleted soil.

We Europeans arrived with steel axes — and firearms. The English elites saw a great opportunity and set forth to dispossess the current inhabitants, to clear the old growth, and establish plantation agriculture. However, eastern North America is not the agricultural equal of the Mediterranean. The east coast of what is now the United States is not as fertile as even the dryish east of Britain, where relatively unleached soils produced a healthy, vigorous people. Highly leached soils such as are found in Cornwall and Wales (west Britain), Massachusetts and Virginia naturally grow forests; but such soils do not produce high levels of health and vigor in people trying to grow grain or livestock on them.

Leaching has more to do with soil fertility than any other factor. The amount of soil leaching determines a crop's nutrient-density unless the grower can wisely import plant nutrients and reverse its effects. If you want to understand your own soil, you must take leaching (or lack of it) into account. Soil starts out as rock fragments that eventually weather down to nothing. Every grain of soil that stays in place long enough, will eventually dissolve into the soil solution.

Dissolved minerals can be grabbed by plants before they leach out. When these plants die (or are eaten), their bodies fall to earth (or the animal's manure does — or the animal itself, when it dies). Then these organic materials decompose back into the earth from whence they originated; the minerals in that organic matter are again released into the soil solution, and new plants have an opportunity to assimilate them. This whole process is referred to as "the carbon cycle."

Full decomposition of woody forest materials back into the simple mineral elements they originated from can take several hundred years.

Consequently, undecomposed organic matter thickly accumulates on the forest floor in the form of *duff* and on prairies as something like lawn thatch, called *sod*. The organic movement sees this process as an example of The Law Of Return — in nature, everything taken out of the soil by plants is ultimately returned to that soil, and rather more marvelously than merely returned, it is often returned in a form that strongly resists leaching, perhaps for centuries. If plant nutrients were not stored in organic matter, the entire planet would support much less biology than it does now. It seems to me as though there is Intelligence doing everything possible to allow Life to build up to the highest possible level.

North America was in the state of equilibrium just described before its old-growth forests were cleared and its native prairie grasses plowed in. When colonists cleared the trees, they plowed in the spongy, mineral-laden layer of partly decomposed organic matter that had accumulated over centuries. In its natural, shady, undisturbed position, forest duff decomposes slowly. Immediately below the duff is a thin layer of dark-colored soil that is rich in humus and minerals. But continue on down an inch or so beneath that layer, and the soil usually is not rich; it is obviously leached. Now, remove the trees, expose the soil to the sun, and plow that inch or two of rich forest duff into two million pounds of relatively infertile soil per acre. The huge savings account of several hundred (or thousand) years of soil mineral accumulation starts decomposing rapidly, releasing its nutrient load. The immediate result is a harvest of bounteous, nutrient-dense crops — for a few years. But then those crops decline. Does that story sound familiar? Lately, we've been hearing of similar goings-on in the Amazon rainforests.

Not every farm newly wrested from forest declines at the same rapid rate. Leached soils that derive from highly mineralized rocks may release enough new nutrients every year that, with good management, good food can be grown indefinitely. But soils like this are rare. Consequently, the agricultural history of the eastern United States resembles that of today's Amazon: it is a story of temporary exploitation, sort of large-scale slash-and-burn agriculture. New lands were cleared; small pioneering farms thrived for a few years and then declined. Livestock started declining, too. So did human health. Many folks moved further west to build new farms and repeat the cycle. As a schoolboy, I recall reading the archetypical American story of this sort: Abraham Lincoln's biography. When Lincoln was a small child, his family exhausted one clearing on the Ohio frontier; then they exhausted a farm in the relatively richer Indiana; and then they moved again, to the fabulous riches of the Illinois prairies.

Thus, you see how the pine forests of the Southeastern states came to be. The soil was mineral-poor soil to begin with. After clearing, good crops grew for some years — until the soil was exhausted. So, the farmland was abandoned. This land, often deeply gullied and washed of all topsoil, then grew another sort of far less useful scrubby forest. Similarly, Genesee, New York, once known as the "flour city" (for the high-quality bread wheat once produced in the region) is now known as the "flower city" because high-protein wheat won't grow there.

I hope those aspiring to buy a country homestead keep this unpleasant history in mind. If you are living on or about to buy a piece of rural land located east of the 98th meridian, and if the land was once a farm that is still clear of forest regrowth, most likely that land has been thoroughly mined out. If the land you're considering has youngish trees that you plan to clear to make room for a house and garden, keep in mind that if it once was a farm field, it already has been thoroughly exhausted. A mere few decades in a second-growth forest will not restore the nutrient reserve that land once carried in its surface humus. A few hundred years might. A few thousand definitely will.

The settlement of Tasmania where I live now was more or less the same story. Around 1810, there only were a couple of villages located at good anchorages. People then moved into the interior, opening farms. After running out of untouched grassy parklands, they started on Tassie's forests. Where the trees grew their tallest and had the hugest trunks, they knew the soil was deep, moisture retentive and likely more fertile. Naturally, these were the first forest lands cleared. Today these fields are still in production and should remain productive as long as there is fertilizer. Where the forests were less lush, the trees smaller, the topsoil thinner, the land sloping, and sometimes a bit stony, fields produced cereal crops for some years, but when exhausted become grazing lands and fruit orchards. These days, the only places you still find old-growth forests are inaccessible rocky hillsides. My point: do not expect that exhausted land will ever grow nutrient-dense crops unless it is first remineralized.

Pat Coleby's book, *Natural Farming,* briefly tells the story of William Evans, a Welshman who came to Australia early in the 19th century, settling in central Victoria where he developed a wheat farm after first clearing the forest.

> *Warden wheat was a variety then popular in England because it had good straw, suitable both for fodder and thatching. In Australia Evans reported, it grew 2.3 metres [7½ feet] high the first year. The farmers, forgetting how they had laboriously maintained soil fertility in their home countries, thought they had struck agricultural gold, and they replanted again — and again. After eight years of monocropping, Evans observed the wheat struggled to reach a height of 20 centimeters [8 inches]. It was by then trying to grow on earth denuded of all organic matter and, as we now know, on basic soil which has never carried the necessary lime minerals. The Welshman bemoaned the fact that none of the farmers put back anything into the soil as they had all done in their countries of origin... They were deluded at first by the apparent great fertility. Even now this attitude persists."*
>
> (p. 7)

After the War Between the States, Americans plowed up the prairies. On these naturally fertile soils, agriculture initially went better. Fifty years later, these soils were falling apart. Prairie farming started out being highly profitable. Prairie soils get just enough rainfall to grow cereals, so they were less leached and far more highly mineralized. To make it better yet, the basic stuff these soils were made from was mineral-rich dust, blown in from the dry lands bordering the Rocky Mountains. This dust still blows in and accumulates at a slow rate. Thousands of years of growing grasses had built a thick, nutrient-rich sod that decomposed rapidly when first plowed in, resulting in incredibly large yields of the highest quality — no fertilization needed. The export of bread wheat grown on these prairies, its handling, transport and finance, as well as transport of supplies to the farmers, created much of the wealth of American elite families.

Leaching and Evapotranspiration

Leaching, the downward movement of rainwater through the topsoil, through the subsoil, and ultimately into the groundwater, steadily and relentlessly removes soil minerals. In humid climates, leaching, all by itself, even without the assistance of erosion, will dissolve all the minerals present in a soil, given enough time. Soil can be viewed as a complex living organism comprised of tens of thousands of interacting species, most of them collaborating to slow the land's inevitable destruction. To do this, Life converts dissolved minerals into relatively stable biomass that accumulates at the surface, where it will slowly decompose and where its decomposition products can be immediately captured by other living things and not disappear into the groundwater. ☛

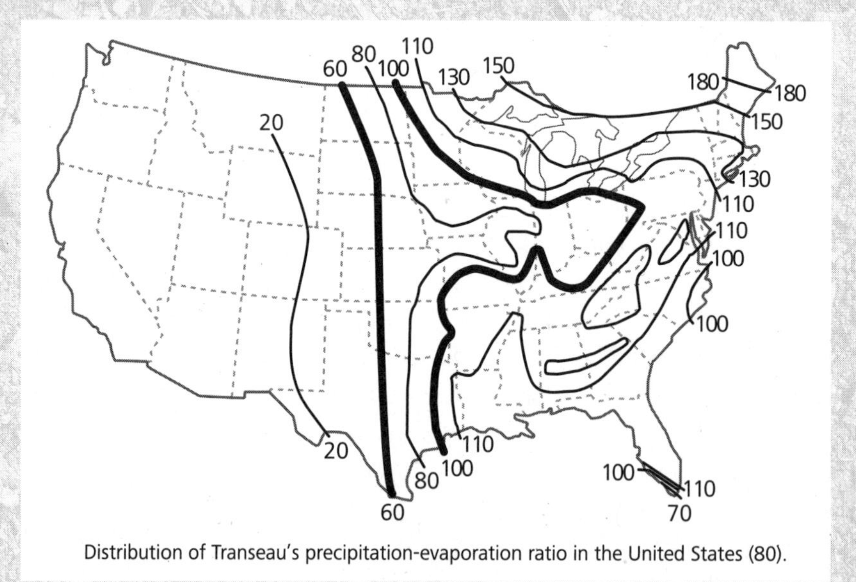

Distribution of Transeau's precipitation-evaporation ratio in the United States (80).

Fig. 2.1: *Evapotranspiration in USA.*

The evapotranspiration ratios mapped here show the average amount of moisture entering the soil from rainfall each year compared to the average annual amount that is withdrawn by sun, wind and plants. Where the ratio exceeds 100 over a year's time, more water goes into the soil than evaporates from that soil and/or is transpired from plants (similar to evaporation). The more the ratio exceeds 100, the greater the amount of moisture that moves downward through the soil, and, all things being equal, the more leaching there will be. Where the ratio is less than 100, rain enters the soil but goes in less deeply; leaching happens only in years of unusually high rainfall. Where the ratio is below 80, soil moisture is usually withdrawn by plants and transpired through their leaves before enough of it accumulates in the soil to start moving nutrients deeper than the plants can draw them back up. Where the ratio is above 100, the natural vegetation is forest. Between 60 and 80, the predominant native vegetation is primarily tall, lush grasses. As the ratio drops below 60, the land is ever-more sparsely covered by ever-shorter grasses. I have heard Texans comment that east of a north-south line running through Dallas, Texas, the land is green; west of that line, the vegetation is usually browned off. This general circumstance continues northward into Canada.

Note that the map is incomplete; there should be another "100" line running down the ridge of the Cascade Mountains of Oregon and Washington, extending into British Columbia. West of that line, the ratio exceeds 100 by quite a bit, except for droughty southern Oregon and a few small areas in the immediate rain shadows of either the Olympic Mountains or Vancouver Island. In the Cascadia bioregion, the leaching from 40–80 inches of rain each year is made worse than it might be because almost all the rainfall occurs during winter, a season when there is next to no evapotranspiration.

However, prairie farming was a form of soil mining. As the land became increasingly depleted, the quality and amount of the grain declined, and then the soil started to blow away.

The prairies regularly experience drought — long weather cycles involving multi-year periods of little rainfall. Although grasses die during a severe drought, their dense sod holds the soil in place until the rains return. However, should the land dry out completely without being protected by a perennial grass cover, the soil blows away, especially when that soil has been exhausted first. During the drought of the 1930s, the winds brought thick dustclouds all the way from Nebraska to Washington, D.C., darkening the sky at midday.

I hope you don't mind that I take a moment from the flow of this conversation to mention Wes Jackson. If you never heard of Wes, you should'a. He is a non-institutionalizable, independent, fully and formally qualified philosopher-scientist who created a private agricultural research institution and upper-level educational facility in Salina, Kansas, called The Land Institute. The Institute's purpose is to work out methods of growing perennial cereal crops on the prairies instead of the annuals we depend on now. The idea is, once a perennial grain crop gets established, it will yield for seven or more years before the field has to be reestablished or the land converted to growing something else. A perennial seed crop means that plowing and replanting need be done only during peak rain years, so that within a few months of being reseeded, the field would again be well covered with vegetation. In times of drought, the soil would be protected by a thick sod.

The 98th Meridian

From the 1930s until about 1960, Dr. William Albrecht served as Head of the Soils Department at the University of Missouri. During Albrecht's tenure, his department produced an incredible amount of useful wisdom, not merely academic gobbledegook. Albrecht's experiments revealed precisely how patterns of soil fertility determine animal (and human) health. He taught methods for managing farm (and garden) soils so they would produce the best nutrition.

I suppose MU's senior management must have been brave to retain Albrecht, tenured or not. He was vilified by a self-serving fertilizer industry; his publications were rejected by most university agronomists. In my opinion, the reason academics opposed Albrecht was because professors who wanted to advance their careers had to please the interest groups and foundations that provided grant money. When you follow the serious money, you arrive at the major agricultural chemical and fertilizer businesses.

In Albrecht's day, official ag establishment farming guidelines, the universities, and the government all asserted that farming required only NPK (nitrogen, phosphorus and potassium) fertilizer — and sometimes lime to adjust soil pH. Soil organic matter was of little concern, and, in any case, deeply indebted farmers could not afford to do what it takes to rebuild soil organic matter. So it became the duty of the agricultural extension people to help figure out a way to grow crops anyway. The party line never linked animal disease with soil quality. Instead, diseases required the appropriate remedy, if there be one, or vaccine, if there be one. And if there was no cure, Authority had an excuse to quarantine and then cull livestock by the tens of thousands, and, as Chuck Walters (publisher/founder of Acres, U.S.A.) put it, thereby creating shortages, thereby destablilizing farm-gate prices, thereby forcing farmers to become gamblers that must borrow ever-more money to survive, putting them ever-more under the control of their banksters. If Walters's assertion seems cynical, I point out that many livestock diseases are entirely preventable by establishing good nutrition. And many are curable using the same approach. Vaccination? Ridiculous practice!

Albrecht is rarely mentioned in garden writing. Gardeners have become accustomed to pabulum — pre-chewed, half-digested (and often half-baked) ideas that take little effort to assimilate and little consideration to apply. But Albrecht's dense writing style requires readers exercise their vocabulary and have a willingness to pause to contemplate the full meaning and implications of his concise statements. In other words, reading Albrecht means doing a bit of hard mental work. And it helps greatly if the reader still vaguely remembers high school chemistry and biology.

Albrecht managed to attract a great deal of opposition. After its first few years, Rodale's *Organic Gardening* magazine never again mentioned Albrecht. I think it's because he didn't pander to J.I.'s moralistic prohibition against synthetic chemical fertilizers. Albrecht's work supports the belief that disease and insect problems are rarely seen if due attention is paid to soil fertility. This did not endear him to the makers of disease and insect remedies. I suspect Albrecht also attracted a great deal of covert opposition from the medical industry. I opine that a few of the chiefliest chiefs around the AMA and/or the University of Chicago Medical School knew they had a lucrative business going and did not wish other doctors or the general public to learn that patterns of soil fertility actually create human health or disease; that sickness is rarely caused by "bad" bacteria or "bad" genes; or that the fundamental treatment for human (and animal) disease is not medicine, but better farming.

Considering the suppression that has landed on William A. Albrecht's message, I'm making the assumption that Albrecht is unknown to you, and that even if you've seen his name in print, you have never actually read him. So I am going to do my best to catch you up with a few of Dr. Albrecht's essential ideas. Albrecht's one actual book (most of his publications were in journals) *Soil Fertility and Animal Health,* is available online for free download. I hope you'll read it. Nah...I hope you'll buy it in hardcover and shelve it next to Weston Price's *Nutrition and Physical Degeneration.*

Albrecht discovered that the most fundamental thing controlling nutrient-density is the evapotranspiration ratio. The more leaching there is, the fewer the remaining soil minerals, and the less nutrient-dense the food — and the balance of minerals in that lower-quality soil shifts unfavorably. On little-leached short-grass Midwestern prairies, it was normal to find in excess of 50,000 pounds of elemental calcium in a furrow slice (2.5% by weight), whereas in the humid Southeast,

Location	N	P	K	Ca	Total**
Minnesota	1,200	3,600	8,400	70,000	82,000
Ohio	4,400	2,000	90,000	18,000	110,000
North Carolina	2,800	5,200	22,000	12,000	39,200
Georgia	2,000	800	2,400	14,000	17,200
Western Missouri	5,600	11,600	58,000	34,800	104,400
South Dakota	11,800	72,000	76,000	55,000	203,000
Nebraska	2,800	48,000	113,000	70,000	231,000
Arizona	800	68,000	106,000	91,000	265,000
Willamette Valley*	2,300	2,200	48,000	15,000	65,200

* Amounts for the Willamette are guestimates worked out by me in the late 1980s from study of an OSU publication summarizing hundreds of Oregon soil test results.
**Total mineralization (P + K + Ca) was not in the original chart, but is revealing.
Source: Soils and Men, USDA 1938 Yearbook of Agriculture, p. XXI

Table 2.1: *Number of Pounds of Elements in Top Seven Inches of an Acre of Land.*

12,000–14,000 pounds of calcium per acre would be the usual. Consider the chart shown here from a paper by Albrecht published in *Soils and Men: USDA 1938 Yearbook of Agriculture*. It shows the total number of pounds of several elements present in the top seven inches of an acre of land.

The proportionate weights, especially the balance of calcium to potassium, shift with the climate. Leached soils contain more potassium proportionate to the amounts of calcium (and phosphorus) they hold. You can see this in the chart. The consequence of these differences is that most crops — fodder, forage, cereal, vegetables — show corresponding and extremely meaningful differences in nutritional outcomes.

Albrecht provides detailed comparative crop analyses, amino acid by amino acid, mineral by mineral. They prove indisputably that the nutritional quality of food depends primarily on the mineralization of the soil that it grew in. Moreover, they prove that the evapotranspiration ratio predicts nutritional outcomes.

Leached soil retains relatively more potassium compared to other nutrients; I could about as well say that leached soil loses relatively less potassium than it loses calcium and phosphorus (and magnesium, etc). Consequently, leached soils produce foods that are higher in carbohydrates and lower in protein, and the smaller *quantity* of protein is also lower-*quality* protein. This sort of food provides our bodies with much more potassium than we have any use for, and it has considerably less calcium, magnesium and phosphorus than we need — desperately need.

When lecturing, Albrecht often told stories. Many of them are lessons he learned from observing livestock. The following story concerns humans. To appreciate this one, you first have to consider what the evapotranspiration map says about the state of Missouri. It shows a considerable difference along an imaginary line drawn from the northwest corner of the state to the southeast corner. There is annual rainfall of about 30 inches at St. Joseph in the northwest corner; there is more than double that amount in the Ozarks, which are southeast. I'm sure the soil numbers for Missouri in the chart refer to the northwestern corner. Prior to World War II, the great majority of foods eaten by a great majority of Americans came from farms fewer than 50 miles from their home. Thus, health and disease statistics of the era show profound regional differences. In 1940, preparing for war, the American government instituted universal conscription. All young men between the ages of 17½ and 26 were required to report in for a pre-induction physical to determine their suitability for military service. In the northwest of Missouri, 200 young men per 1,000 were deemed medically unfit for military service; in the southeast of Missouri, 400 per 1,000 were found to be unfit, and from the middle of the state (along that imaginary line), where rainfall was about half of the extremes, 300 per 1,000 were unfit.

Similarly, the Army at that time collected statistics on dental health; the number of cavities in inductees was directly related to the evapotranspiration map, showing great differences in dental health depending on the soil. These days, when most people eat from supermarkets that bring food from everywhere, such differences have largely evened out — for the worse. And since all industrial farming is done using much the same economically rational soil-fertility management approach, our average health these days more resembles that of people from the Ozarks in 1940 than the people living around St. Joe.

The feeding studies Albrecht did with rabbits provide further indisputable proof. He fed five identical groups of rabbits on five different lots of hay brought from each of the four corners of Missouri, and one from the center of the state. On different feeds, the groups of rabbits entirely changed appearance and overall health. The bunnies that fed on hay from northwest Missouri became half again larger, bred well and lived long. The bunnies on hay from the Ozarks shrank; they became about half the size and weight of those fed on northwestern Missouri hay, had much shorter life spans, and when autopsied, their livers did not look so good. The rabbits fed on hay from the state's center were intermediate in size, health and longevity.

Albrecht concluded that we should target the typical mineral ratios in prairie farm soils as an ideal balance of soil calcium, magnesium, potassium and sodium that produces the highest possible nutritional outcomes as well as bountiful harvests. His is not a precise prescription but it does, at least, distinguish the target. Following Albrecht,

Fig. 2.2.

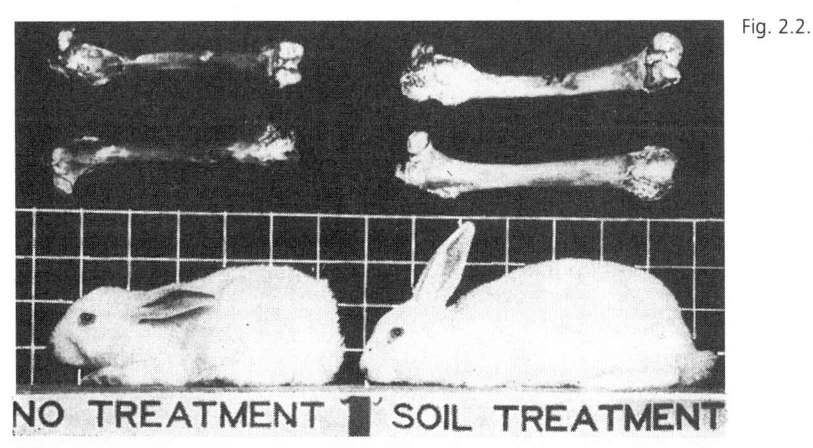

Fig. 3-IV. The weanling rabbits had the same pedigree, so did the crop plants making up the hay, but treatment of the soil with some extra fertility to grow better feed made the rabbit on the right different in appearance and body structure, as the bones also illustrate.

to create nutrient-density, you first measure the amounts of minerals present in soil and then adjust their balance toward a predetermined target. When soil nutrient levels approach the ratios that Albrecht was talking about, you get much better food.

The Astera system, described briefly in the Introduction, refines Albrecht's insights to the point where ordinary gardeners can analyze their own soil test results and thereby nudge their own soil-fertility pattern in the direction that will create peak nutrition in the food being grown. Because so many people have been handed oversimplified, incorrect ideas about soil fertility and about what organics is all about, I say it again: *soil mineral balancing is not in any way a contradiction of the organic system; it is a natural extension of it as well as being the prerequisite action needed to make organics work as well as they should.*

Chapter 3

The Shit Method of Agriculture (SaMOA)

Soils vary greatly in their potential to produce nutrition. This potential determined by the nature of the parent rocks, but even more so by climate and rainfall. Climax ecologies in temperate humid (forests) and sub-humid (prairies) climates build up large quantities of decaying biomass that serves to horde previously released mineral nutrients. That's why just cleared old-growth forests produce nutrient-dense food for some years, and prairie soils can produce heavily for decades after they are first plowed. Americans, Canadians and Australians all enjoyed this huge advantage — while it lasted.

After the soil's nutrient bank account has been drawn down below a critical level, the soil must be remineralized if productive agriculture is to continue. Humans have long attempted to do this by applying animal (and human) manure and, since Roman times, lime. If there's one topic that deeply interests me, it's how to farm so as to grow nutrient-dense food. If there's one topic I am not qualified to teach, it's how to farm — because I have never done it. Still, it is useful for gardeners to contemplate farming, as long as we do not unthinkingly imitate what works in farming.

The Soil and Health

Sir Albert Howard's classic book *The Soil and Health*, tells us what happens in humid temperate climates when people farm but cannot (or do not) fertilize. It happened after civilization broke down in the Western Roman Empire and Europe moved into a dark age of turmoil, war, pillage and generally dangerous conditions. A security arrangement developed that we now term the feudal system, which, at least in England, had an associated method of farming called *the strip system*, or the *two- or three-field system*. It is difficult to envision a poorer agricultural system.

Villagers farmed in semi-collectivist fashion. After the commonly owned arable fields had been plowed by common effort, they were marked out into long, narrow strips, usually 10 yards wide by one furrowlong — 220 yards. Strips were assigned by lottery; each family could get up to 30 of these half-acre-sized strips to care for and harvest. Their strips would be scattered at random around the large field to equalize better areas from poorer ones, and strips were assigned only for one year.

Every year, grain production was rotated to another of the village's two or three open fields, thus the "two-field system" or the "three-field system." To keep weeds under control, the resting field was plowed repeatedly, so kept entirely bare, which meant its organic matter level declined even further. There being little chance of getting the same strips in subsequent years, there was no reason to improve the strips your family had been assigned for that year. Besides, manure was scarce; the villagers' scabby livestock were kept on a severely overgrazed commons. There being no community land to use for growing hay, there was no way to accumulate manure, and generally, what little manure could be gathered was the property of the lord of the district, to be used on his fields. In most places, the fertility of grain fields and pasturages declined steadily for centuries until it stably bottomed out.

Fertility did not go to zero, because every year a small fraction of the soil particles in that field dissolved and released their mineral content. Albert Howard called this release "The Annual Increment of Fertility"; it provided all the nutrients those crops got. By today's standards, yields were absurdly low. Two bushels of grain had to be sown per acre (they had no grain drills at that time; all sowing was done by broadcasting, an inefficient method) to harvest, maybe, eight bushels. I imagine the nutrient-undense wheat berries of that era were tiny and germinated weakly, another reason two bushels per acre were required. As the soil settled into its lowest ebb, people became ever-more poorly nourished. Then in the mid-1300s, Nature put things back into balance; a series of plagues came that took away over half the population.

Everyone in Europe was exposed to this virulent infection, but the disease entirely depopulated some districts while barely touching others. Historians say the Black Death took 60% of the total population of Europe, which is probably a reasonably accurate statistic. But considering only the percentage gives a distorted picture. A more informative one would show the pattern of plague deaths compared to a map of fundamental soil fertility. A statistical study comparing parish records against already existing soil mapping would make an excellent doctoral dissertation. But even without academically acceptable proof, I already am convinced: in those places where the underlying rocks provided a larger annual increment of fertility — where, I would guess, the wheat harvest at that time would amount to twelve bushels per acre — there probably were few or no plague deaths. On poor ground, entire communities vanished, and these soils went back to forest. Farming stopped in many areas for nearly a century, giving the weaker European soils a chance to accumulate some organic matter again and to build up fertility in the same way that virgin land does.

During the Renaissance, farming improved hugely. Scientific agriculture began to interest the European elites; being a successful estate manager became at least as fashionable as the old aristocratic standbys — extortion, pillage, war and sensual excess. Great agricultural improvements came from this new interest. The beneficial effects of green manure crops were rediscovered (the Romans had used them). Multi-year crop rotation patterns were instituted (again, the Romans had already done it). Liming was again practiced. And careful breeding was applied to both crops and livestock. Consequently, yields increased; animal and human health improved.

Then came the Industrial Revolution. The English peasantry, now required to tend machines, were ruthlessly forced off their traditional holdings and into the cities to work the mills. This population shift also allowed economically rational consolidation of traditional village lands into the hands of the upper classes; open fields became large enclosed blocks that could be farmed profitably in a businesslike manner. The overall food supply increased in nutrient-density, so the average health and vigor of the English improved. With this newfound energy still under the direction of a predatory aristocracy, the British proceeded to conquer much of the planet.

The SaMOA Method

All this came from intelligently applying what I once jokingly termed the Shit Method of Agriculture ("SaMOA," pronounced like the tropical South Pacific island group). I apologize if my use of the term "shit" is uncomfortable for some, but the acronym SaMOA leads to such a good pun that I can't resist. The SaMOA grower fertilizes with composted organic wastes, composted and fresh animal manures, and, in some places, humanure. Dung can be spread fresh on the field and left to itself; this practice gives some benefit, despite huge nitrogen losses. Better, fresh manure can be worked into the soil so that it decomposes more efficiently, thus improving the soil more effectively. Manure can be gathered, cherished, guarded against leaching, attentively composted in heaps, and then spread and worked in. This is best.

The more scientific use of poop and green cropping, which increased the overall amount of organic matter and soil nitrogen, made all the difference to European farming. It steadily developed along these holistic lines from the 1600s until the later 19th century.

A limited number of artificial chemical fertilizers appeared in the mid 19th century, but were little used. Most farmers of that era considered artificials too costly for the benefits they gave. However, natural, mined fertilizers such as guano and Chilean (sodium) nitrate were inexpensive and became popular. The First World War prompted the construction of chemical factories to synthesize explosives. After the war, this productive capacity was converted to producing nitrate fertilizer, while the wealthy factory owners spent time and treasure convincing farmers to use their products. Thus began our current system. Industrial farming can produce one or more cash crops on every field, every year. It neglects recycling animal manures which had long been the farmers' source of nitrates (and phosphorus). It neglects growing green crops to keep organic matter and nitrogen levels up because nitrates are cheaper and easier to get from a sack.

Farming became an economically rational business that had little to do with family self-sufficiency. Without animal manuring, and with cash crops steadily going out the farm gate, soil organic matter levels inevitably dropped. Precipitously. Minor nutrients (copper, zinc, boron and manganese) also went out the farm gate. But they were not included in chemical fertilizer blends. In consequence of this depletion, new or rarely-seen-before crop and livestock diseases appeared and quickly became commonplace. Contending with animal disease soon seemed normal. It became ever-more costly to maintain production levels as more and more inputs were needed to counter the loss of soil fertility. Traditional varieties that were once highly productive "ran out," meaning seed weakened to the point where it could no longer establish a fast-growing stand in the field. The real reason: steady depletion of soil minerals and soil organic matter. The official story: varieties run out; science will breed newer, better ones.

The organic farming and gardening movement began as an attempt to restore health to the farming system and to those who ate its food. Starting in the 1930s, it called for a return to tradition. It pointed out essential, scientific principles that had been neglected in the rush to profit from cheap chemical nitrates, mainly, The Law of Return — mineral nutrients taken out the farm gate must be returned to the soil; organic matter must be maintained or its level will drop, and the soil ecology will start to die off.

Rodale

J.I. Rodale's *Organic Gardening and Farming* (OGF) magazine introduced organics to North Americans. Over-the-top with earnest, almost revival-tent enthusiasm, *OGF* preached Rodale's take on the SaMOA method; approaches that differed from the Rodale Organic Doctrine were not mentioned; in the early years, they were directly criticized. I started gardening as an *OGF* subscriber. I closely studied every issue from 1972 to the mid-1980s. A decade later, I spent a long afternoon at the OSU library in Corvallis, Oregon, critically reading the early issues of *OGF* (1945–50) in an attempt to heal the damage caused by my earlier, uncritical acceptance of that magazine's belief system.

I can fairly summarize the essential aspects of Rodale's approach in two sentences: To grow an abundance of highly nutritious vegetables and fruit, make and then dig in compost. Lots of it.

We were repeatedly told that the successful organic gardener must import heaps of organic waste and then compost it before feeding it to the soil. Or else, you could spread that waste thickly over the soil and then shallowly till it in, letting it *sheet compost*. The magazine urged us to obtain organic wastes from wherever they could be found for the hauling, especially around one's own neighborhood. Bring the organic matter level of your garden ever upward. Bring this black gold home! (So I bought a pickup truck.) And then, if the soil is acidic, counteract that undesirable condition by adding crushed limestone to bring its pH close to neutral.

According to Rodale, soil acidity is "bad," and measuring it is easy and cheaply done. J.I. often said if you're going to add lime, it is better to use a sort called dolomite because this type contains both magnesium and calcium — and magnesium is as much a vital plant nutrient as calcium is.

And that about summarizes the organic system's essentials according to J.I. Rodale. His magazine and book publishing company taught several generations of gardeners that it takes manure, compost and lime to grow a great garden that will make your health rapidly improve — because your food will become as nutritious as food can possibly be.

If, along the way, something did not grow all that well, the solution was either to add more lime if the soil pH was still too acidic, or to add more compost — or better compost. Usually the choice was to add more compost. And that is why I put that mysterious lower case "a" in the acronym SaMOA. Because what the Rodale Organic Doctrine essentially comes down to is this: if it don't grow well enough, then just add SaMOA.

This method has great instinctual appeal. It matches the human genome imprinted by tens of thousands of years of surviving through raising food that way. Baby not happy, give it some milk. Plant doesn't look good; give it SaMOA water. Or SaMOA something.

About those rewards: A new organic food garden usually grows great for the first few years. The gardener starts out by digging in compost and/or manure. These decompose, supplying mineral nutrients that feed the plants while the carbohydrates in this decomposing organic matter fuel a rapidly multiplying soil ecology. Increased soil microbial activity releases the soil's existing mineral nutrient reserve more rapidly. Consequently, the crop gets a double dose of nutrients — decomposing compost plus enhanced nutrient release. So the garden grows well, just like the organic literature predicts. In humid parts of North America, lime is routinely added; lime releases calcium into the soil, and magnesium too, if dolomite lime is used.

After a few years, heretofore unknown and/or previously unexperienced diseases and/or insect attacks usually arrive. The usual explanation offered by the local ag authorities is that it took a few years for the insect or disease to stumble into your garden, but now that it is present, there's no getting rid of it. This actually is possible, and the certainty of the ag agent might make it seem probable. The solution proffered by your local extension agent usually requires repeated spraying. The actual solution, the one that eliminates the problem instead of fighting it, is beyond most of them.

The organic solution is to bring the soil to an even higher level of fertility by digging in some more compost or maybe mulching with it. Actually, that answer goes in the right direction, but it is still the wrong answer. Yes, the soil needs to be brought to a higher level of fertility, and yes, properly nourished plants are usually little damaged by insect or disease; this is really true, but ordinary compost and a bit of lime are rarely what is needed to achieve sufficient nourishment.

Vegetables grow poorly in tight, airless soil. Organic gardeners lighten soil by mixing a great deal of compost into it. Farmers without debts can grow green manure crops. Either action ups the soil organic matter level, which almost instantly transforms soil. The expression gardeners typically use for this transformation is "building up the garden," because when you increase the soil's humus level, the earth actually does lift itself up, because it contains more air than it did before. (I have seen my own beds literally elevate themselves several inches as a crop of strong pasture grasses filled them with roots.) But soil compaction can be hard to conquer in some organic gardens. Sometimes, no matter how much compost is put in, the soil doesn't seem to stay loose. This seems mysterious because, according to the Rodale Organic Doctrine, organic matter effectively lightens compacted soil.

J.I. Rodale's publications repeatedly expressed a strong preference for dolomite lime. Pound for pound, dolomite lime raises pH more than high-calcium lime does. So, by using dolomite you seem to get SaMOA for the same effort and cost. However, high magnesium levels change the behavior of clay, making it want to stick to itself and pull itself together into an airless, hard, compact mass. The organic gardener, surprised that additions of dolomite lime and compost have not sufficiently loosened the soil, usually assumes that even larger quantities of organic matter are what is needed.

But compost rarely contains the ideal mineral balance to grow nutrient-dense food. Excessive additions of compost usually imbalance the soil's mineral profile and degrade nutritional outcomes. William Albrecht clearly explains how this works, but when I started gardening, no garden writer that I read said anything about Albrecht. Thanks for that, J.I.! Whenever the gardener of my era consulted *OGF* or the countless how-to-garden books in print at that time, the answer usually was to add more compost and lime, preferably dolomite.

Here's the truth: If too much high-magnesium lime gets added to a soil with clay in it, compaction, airlessness and tightness can increase — despite huge additions of compost. Magnesium excess can tighten up a clay subsoil beneath a sandy topsoil, preventing the crop from putting roots there. Magnesium's effect is amazingly powerful when the soil has a lot of clay in it, even ten percent clay and too much magnesium can make soil become rock hard and airless, even if it has had heaps of organic matter put into it.

Gardening books in the tradition of the Rodale Organic Doctrine (ROD), which is almost all of them after 1950, suggest that when there are enough worms, when the humus level of the soil has gotten high enough, when the soil has been limed so its pH is between 6.2 and 6.5, then everything will work sweetly. These books conveniently overlooked the many people who had already gone down this path and, after four to six years of growing by the organic book, found the garden they were so proud of began to grow poorly. And it wasn't making them healthy.

Perhaps you think I'm confusing my own imagination, or my own unique experiences with what happens to most people. However, when I became a volunteer lecturer on organic gardening for the OSU/WSU joint Master Gardener training program, I talked to a lot of people about what happened with my teeth and overall health after eating too many vegetables from my organically grown, unbalanced trials ground. I had people bravely come up after my talk to tell me that their health had similarly deteriorated when they depended heavily on their own organically grown food. They thanked me for clearing up the mystery.

Here's what happened to me and my family: initially, the biologically enlivened soil did a great job of making mineral nutrients available. But, after a few years of heavy withdrawals, some of these nutrient elements declined below a critical concentration. This might never have happened if *everything* taken out of the garden had been returned to it, but few gardeners return their family's urine and humanure. Organic gardening literature implies, and sometimes directly asserts, a sub-rule of The Law Of Return, I call it the "Law of Equivalence." It says that imported compost and manure should contain all the plant nutrients we sent down the toilet. This assertion is almost inevitably incorrect, although there are some extremely fertile regions in North America where the Law of Equivalence applies to local compost and manure.

The typical organic garden has usually been limed frequently, so it contains excess calcium, and usually, excess magnesium if dolomite was used. It will have an extraordinarily high level of organic matter, so there probably will be plenty of nitrate nitrogen present, or at least in the heat of summer there will be. But crippling deficiencies (or damaging excesses) of other vital nutrients have probably developed. I expect imbalances develop differently in every part of the continent, depending on climate and the soil's parent rocks. The area I know best is Cascadia. There, the soil holds huge supplies of potassium but insufficient calcium and magnesium to properly balance that potassium; these soils are also typically short on phosphorus. Willamette Valley soils are ideal for growing low-protein, nutrient-undense soft white wheat. Upland soils are perfect for non-food-producing tree crops, like Douglas fir. Plants inevitably concentrate potassium into their structural parts, so enormous quantities of potassium are brought into a typical Cascadian home garden from imported grass clippings, grain straw, spoiled hay and forest-industry wastes. Excess soil potassium makes plants *seem* to grow great, but it also has a devastating effect on the nutritional qualities of that food. More about potassium, to come.

Garden Writers

There are oft-repeated errors in veggie gardening books, and an even denser concentration are found in gardening magazine articles. After all, we garden writers are only ordinary humans — with all the usual flaws and warts — who passionately share your interest in food growing.

Garden writers are well meaning; they try to help others succeed by sharing their own successes. I do not usually sense obsessive egotism distorting garden writing, nor do I strongly smell "I'm better than you are," seeping out the cracks in gardening magazines. I suppose that's because there's little about any home gardening opinion that can threaten anyone else's prosperity or survival. Most garden writers positively push what they believe succeeds, while ignoring those who disagree. That's been my approach too, until writing this book. In this book, I try to broaden the reader's opinion about what organic is and what it should be — and that can be a touchy minefield. Since I'm setting out to change opinions, it seems sensible to help lighten your load of previously acquired views by pointing out their source.

Few garden writers received a formal horticultural education. That's positive in a way, because those who graduate from a university, especially with an advanced degree, usually have been so traumatized from having to produce professor-pleasing papers that rarely are they ever again able to write freely. I think it's near-impossible for someone with an MSHort to write for the home-garden market. On the other hand, garden writers are glib, but they lack proper scientific foundations. They learn from their own experience and/or from the garden books and magazine articles written by other garden writers who have similar limitations. Having avoided the pitfalls of academic horticulture, they lack the formal science needed to appreciate the common flaws found in gardening information. Thus, errors get uncritically transmitted from one generation of garden writer to the next. And with each retelling, the error acquires more authority, seems more correct, because it has been said by Everybody Else so many times for so many years.

Garden writers usually lack broad experience. They typically start out being over the top about growing backyard vegetables and consider they have achieved enormous success. But these great successes can be pretty subjective and actual results quite ordinary because they are comparing their results with their neighbors'. I did that. In my first few books, I enthusiastically, and with great sincerity, encouraged my readers to "do as I have done, and you'll have as great a success in your backyard as I have had in my own." But then I saw Dr. Jim Baggett's OSU variety trials expertly grown on Class I soil. I realized I had a long way to go before I could match those results.

A garden writer usually gardens on one type of soil, in one climatic zone, and perhaps in a uniquely favorable (or particularly challenging) microclimate within that general climate. Contributing factors, such as the salt levels in irrigation water, or a particularly favorable backyard microclimate often go unrecognized, or the gardener just ignores them in the flush of triumph. I did that. When I began growing variety trials, I'd only been gardening for six years, four of them in southern California on pretty good soils. Oregon was entirely different. At that time, I reckoned that growing trial gardens on lousy soil in a short-season, chilly coast range valley was a plus for my business. Although it took a lot of inputs to make that silty-clay subsoil grow things acceptably well, I was able to eliminate the weakly rooting or chill-intolerant varieties that didn't thrive. Since most gardeners don't have high-quality agricultural ground or live in banana belts, varieties capable of performing in my trials gardens would do well anywhere in Cascadia if given half a chance. But my how-to-garden opinions were distorted in those days by the enormous amounts of inputs it took me to grow decent crops. Consequently, my books from those years advised the overuse of seedmeal and the use of far more organic matter than I now consider necessary, even on silty-clay subsoils. In fact, if you really want to embarrass me, get one of my early garden books and quote back to me some of my viewpoints from that era.

Organic Gardening magazine educated my cadre as we came of age. And my group has been educating everyone younger. The dogma underpinning what North American garden writers have long been promoting is best appreciated by reading J.I. Rodale himself. His books, *Pay Dirt, The Organic Front* and *The Healthy Hunzas,* are slim and easy to read, and I've put them on the Internet for free download. If you fancy yourself an organic grower, I urge you to have a read and see where some of your opinions originated from.

Holistic Farming Writers

It especially helps serious food gardeners (by "serious," I mean gardeners who grow a substantial portion of their family's food as much of the year as possible) to distinguish between home gardening/small-scale market gardening methods and holistic farming practices. The first cadre of organic garden writers — the group active during the 1940s and early 50s — took great interest in the many how-to-*farm*- organically books coming out at that time. However, there is a fundamental difference between gardening and farming. To quote Robert Parnes, "the vegetable garden is an endless sinkhole for plant nutrients." It has to be. Many kinds of vegetables require extraordinarily rich soil to grow well. Vegetables were like that during Roman times, and they still are. It's in their genes. But sustainable, biologically oriented farming aimed at producing nutrient-density in field crops operates at lower levels of soil fertility than a garden requires. A farm can export some organic matter; a garden, never.

Sustainable, ethical farming builds soil nitrate and soil organic matter levels by first plowing under a lush legume green manure. However, keep in mind that most kinds of garden vegetables require more nitrate than legume green manuring can usually create. Yet garden writers not equipped with the experience to evaluate farming techniques instruct gardeners to create sufficient soil nitrates by placing bush beans around a nitrate-demanding plant. To pursue that illusory nitrogen, gardeners give up hoeing and switch to painstakingly slow hand-weeding in order to grow bush beans amongst their cabbages, all this to no useful result — just because an early garden writer misunderstood an organic farming book.

With each restatement in each new book, mistakes like this gain credibility. For example, in one of Newman Turner's farming manuals, he brags about building up an infertile pasture that previously could only feed 50 sheep by putting 200 sheep on it so their manure would enrich the soil. Turner briefly mentioned, in an easily overlooked aside, that because there was not sufficient grass for them on this exhausted pasture, these 200 sheep were fed "oilseedcakes," (cottonseed meal), and because the grass was so poor, (high-quality) hay was brought in for them from elsewhere. Sheep only assimilate about a third of the nutrients going into their mouths, so Turner's sheep redistributed the unassimilated two-thirds at random around the pasture, thus remineralizing it. The result on soil fertility was little different than it would have been had Turner spread the seedmeal as fertilizer and composted that hay before spreading it. But many non-farmers incorrectly assumed from Turner's assertion that grazing adds fertility to the soil being grazed. This is especially easy for Americans to do because "oilseedcakes," is a British term that causes North American reader's awareness to momentarily glaze over. I have personally corrected several dozen local hobby farmers blithely passing this one on. The obvious truth is, grazing animals concentrate soil minerals into their flesh and bone, which is then trucked off the farm.

I'm presenting this meditation on the follies of garden writers so that when I make some assertion disagreeing with something you have previously learned — when a viewpoint of mine contradicts an opinion of yours that seems important — maybe you won't immediately reject what I have to say.

The Biggest Mistake

The *Organic Gardening* magazine I learned from made little distinction between soils. Dogma asserted that you could turn any old gravel heap or clay pit into a veritable Garden of Eatin' if only you put enough compost into it. I actually held that opinion myself when I bought those worthless five Oregon acres in 1978. I would not make that mistake again; I should have spent more of our money on land and less on buildings. Any old shack can be upgraded or torn down and replaced. But soil by the acre...what you've got, you're pretty much stuck with. Actually, soils vary greatly in their physical properties, i.e., in their proportions of sand to silt to clay and their depth and slope. And they differ greatly in the amounts of plant nutrients they offer. The most obvious reason for this is differences in the mineral content of the rocks that made the soils. The least obvious reason, and usually the most significant, is differences in climate.

I'm referring to the evapotranspiration ratio again. (Recall the map shown in Figure 2.1.) Some soil nutrients leach more readily than others, so in humid climates, the soil mineral balances fall into a common pattern. Plants respond to that pattern by developing corresponding nutritional qualities. I've already asserted that plants grown on leached soils contain less nutrition than those growing on more fertile ground. Organically grown vegetables fertilized only with compost and animal manure sourced from surrounding soils still reflect the region's pattern. Suppose your district's soils are typically short in one or a few specific elements, say, for this example, sulfur, or phosphorus (or, as in my part of Tasmania, seriously short on zinc and copper and typically low in potassium and phosphorus.) In a deficient district, most of the surrounding vegetation will be similarly deficient in these elements. Fertilizing a garden by composting local vegetation and animal manures derived from that same kind of vegetation will only magnify the regional soil imbalance.

One extremely important soil pattern concerns potassium. If soil potassium gets top-heavy, plants grow differently. Instead of making proteins, they make more carbohydrates. The bottom line is this: crops on high-potassium soils produce about 25% more carbohydrates; at the same time, their protein content is lowered by around 25%. I understand that a number like 25% more, 25 less, might not seem important. But extra calories combined with that much less protein makes your health suffer and shortens your life span significantly; if you are female and of breeding age, your children will not get nearly as good a start — and a human body trying to subsist on this degraded food would constantly nag its owner to feed it far too much for its own good.

Not only does high soil potassium lower overall protein content, the nature of those proteins changes. I hope you already know that proteins are long, complex chains of about 20 different amino acids. A few amino acids usually are scarce; in plants grown with excess potassium these are even scarcer, lowering protein quality and leading to diseases in all the animals eating them, including us. Another shift occurs in the food's mineral content. As soil potassium increases, the mineral content of the plant growing on that soil also shifts. Excessive K (potassium) in the soil results in much higher levels of K in the plant tissues, but correspondingly lower levels of calcium and phosphorus (and minor nutrients). Our bodies can hardly get enough calcium, magnesium and phosphorus, but we do not need high quantities of K. Some, yes; heaps, no.

When livestock eat high-calorie, low-protein, mineral-deficient food, they gain weight readily. But without enough proteins and minerals, they fail to breed successfully and are not healthy. We intentionally fatten livestock on corn — a high carbohydrate, low-protein diet. But when we want animals to reproduce well, to withstand disease, and to live a long time, we feed them differently. It seems to me that if science knows how to make animals fat or thin, capable of reproduction or not, long-lived and healthy, or short, fat and sick — if those big agri-food businesses that manufacture livestock and poultry feeds know how to control the health and longevity of animals, then those running the big human-food businesses, the ones putting all the prepared food products on the supermarket shelves, also know. And without possessing any of their secret business documents, I still know they know because if you follow the money, you'll find the same powerful families control both sorts of food businesses — as well as being big players in pharmaceuticals and medical education (read here, the AMA and the ultra-powerful University of Chicago Medical School). If you doubt me, have a read of Ferdinand Lundberg's classic *The Rich and the Super Rich*, also available for free download at soilandhealth.org.

Perhaps the sorry state of our industrial food supply results from the accidental coming together of many unintended consequences, but I think it's highly possible our industrial food system is *intended* to produce a population that is lacking energy, is low-grade sick most of the time, increasingly stupefied and controllable, and dependent on medicine and doctoring. Like feedlot cattle, Americans clearly have become less successful at reproducing and seem to be more and more dumbed-down. Not to mention, beefy.

What to do? Few soils are fertile enough to grow vegetables without increasing their nutrient levels. But if you bring in fertility by way of local vegetation, either to be composted or having first passed through a cow's gut, you import some nutrients, often in excess, but do not adequately replace others. Thus you further imbalance your food. Unless your soil has developed in one of those rare, remarkable spots where the annual increment of soil nutrient release is huge and complete and balanced, your garden will be increasingly depleted of minerals and/or will develop mineral imbalances; some are minor, but most come with undesirable consequences.

The gardener caught in that dwindling spiral may not even recognize the decline. It may not reveal itself as a catastrophic disease or an insect infestation or a total crop failure; it may sneak up slowly, unnoticed, like nerve deafness. Growers may believe their gardens to be entirely marvelous, but the food they're eating is very probably far from being as nutritious as it could possibly be.

One indicator that soil is going out of balance comes when species that once grew well no longer do. Some types of vegetables seem too difficult or too troublesome to attempt, while those species better able to cope with a garden's soil imbalances will come to be viewed as being easy to grow. Take it from someone who has organically grown commercial vegetable trials — every kind of vegetable, every species, is easy to grow if: 1) the soil is even close to being nutrient-balanced, contains adequate humus, and isn't the densest of dense clay; and 2) they are grown in the appropriate season and are adapted to the climate.

Suppose a region has soil offering large quantities of all essential plant nutrients in close to an ideal balance. North American prairie soils *were* like that. Any manure or spoiled hay or crop waste brought into the garden from this rich, balanced soil will be rich and balanced. Soils like this will not be highly acidic and may rarely, if ever, need lime. In this circumstance, compost gardening can be a great success — indefinitely. Cutting J.I. Rodale every possible bit of slack, I assume that the soil around Emmaus, Pennsylvania — where J.I. had his gentleman's hobby farm on which all his hands-on food growing experience happened during the 1940s — is rich, fertile, balanced soil. Like most garden writers, I am sure the Rodale Organic Doctrines his publications promoted worked as advertised — in his backyard. But can *you* expect a result like that using the Rodale method? Sorry, not so likely. I know only of a couple significant areas in North America where naturally rich soils can be found: the bluegrass country in Kentucky (but not the limestone soils of the Ozarks), and the wheat-belt prairies. I'm sure there are SaMOA rare, rich spots.

So what's a serious gardener to do? You have an answer in your hands.

Chapter 4

Complete Organic Fertilizer

The previous chapter highlighted two main points: 1) Compost gardening grows nutrient-dense vegetables and fruits *when the materials being composted come from balanced fertile soil;* and, 2) When organic matter grown on unbalanced soil is concentrated into a similarly unbalanced soil, it does increase overall mineralization but simultaneously exaggerates the existing lack of balance and does not grow highly nutritious food. Remineralization guided by a soil test is the best method. First, you find out what the actual chemical nature of your soil is. Then you add nutrients calculated to bring that soil into a fertility profile that produces nutrient-dense food. Remineralization also involves building a high level of soil organic matter.

Soil testing and all that goes with it does not match some personalities. Most humans understand things outside the scientific method — through faith, magic, intuition, inspiration, emotion, symbolism. My use of the term "outside" is not judgmental or a put-down. "Outside" does not imply "less"; it could as well mean "beyond." If having a non-scientific personality describes you, using Complete Organic Fertilizer is a parallel approach to soil remineralization that does not require soil testing or precise weighing and spreading of fertilizers. My Complete Organic Fertilizer recipe has a 30-year track record of producing excellent results for tens of thousands of home gardeners. You'll be able to count on it.

Soil testing and all that goes with it does not match some circumstances. A little garden patch growing a few tomato plants does not justify the cost of a test or the purchase of half a dozen ingredients in quantities sufficient to apply half a dozen variations on that original fertilization. A complete organic fertilizer is a better approach, especially if you buy a pre-mixed COF that provides proper balance.

For many years, I believed COF and compost were all the soil-fertility building a food garden needed. Until last year, the COF method formed the core of every garden book I wrote; it's worked far better than compost-only gardening ever since I started recommending it in the mid-1980s. Almost everyone who tries it continues to use it.

The Story of Solomon's Complete Organic Fertilizer

My first garden books were written for a unique climatic region called Cascadia, which includes the northern California redwood country, western Oregon, western Washington, and the Lower Mainland and Islands of British Columbia. Oregon has an informal, folksy culture; local people felt free to bring their questions up my driveway or to ring me about their confusions. So, I was frequently moved to create ever-clearer explanations of even-more-effective even-simpler techniques that were ever-more foolproof. And that is why Complete Organic Fertilizer came to be. Henceforth, I will call it COF to save myself a few thousand keystrokes.

The basic concept was not mine; the idea was handed to me when I was making my first San Fernando Valley garden around 1973. *Organic Gardening and Farming* magazine mentioned a farm advisor named Will Kinney, whose place was not that far away. I reckoned some of the exotic fertilizers he sold would help me, and maybe he could educate me, too. I got lucky; Will liked me. Answered my stupid questions in full. Really extended himself. Maybe he taught me because I really listened. Will knew all about the soils around where I was living. He said my garden was generally well mineralized, but most San Fernando Valley soils lacked the same few elements and of course, he had in stock exactly what I needed — a 50-pound bag of coarsely ground soft, pinkish chunks the size of rock salt. It contained sulfur, magnesium and potassium. It came from New Mexico, where a huge cave system had, over geologic time, become solidly packed with stalactites. My soil needed extra phosphorus. Out came a sack of rock phosphate.

"What do you make compost with?" asked Will. So I told him about the big stables not far from my acre, where I could get all the manure I cared to load and cart away. "Did this manure have any sawdust in it?" asked Will. It did. In fact, it was mostly sawdust, smelling faintly of urine, and not much decomposed at that. "Then you'll need some seedmeal," he said, "to make up for all that sawdust." Out came a sack of cottonseed meal. Then came a long rap about trace elements and how important they can be to getting high nutrient-density, a brag about how he had recently produced a crop of leaf lettuce that "tested at over 20% protein — as good as beefsteak!" And out came a bag of rock dust from Utah that originated as ooze on the seabed floor. The label showed every element I ever heard of, but in quantities of one or two parts per million at best; some were present only as a tiny fraction of one part per million. I can't recall the product name for certain, but it probably was Azomite.

Will instructed me to mix one pint of the pinkish rock (I now know it was langbeinite), one quart of rock phosphate, one pint of the trace mineral rock, and two quarts of cottonseed meal and spread this over 100 square feet. In addition, I was to use all the compost I could make. Will said to spread and dig in another dose of these minerals every time I started a new crop. And thus, without realizing it, I had been gifted with a site-specific Complete Organic Fertilizer. And did my garden ever grow great on it!

Four years later, I was creating a new garden in Cascadia in the Oregon Coast Ranges. I soon discovered that soils west of the Cascades were the opposite of the San Fernando Valley. They are highly leached and rather infertile. The geologic history of Cascadia is such that almost all the rocks in western Oregon and Washington States have similar chemistry; most contain higher-than-usual levels of potassium — much higher. So even though these soils are highly leached, they still contain significant amounts of potassium. The last thing I wanted to do in Oregon was use Will Kinney's formula, which was designed to fortify soils with potassium. (It might help you in reading what follows if you go back to Chapter 2 and have another look at the table of soil mineralization by location. Notice that western Oregon soils have fewer minerals overall; they are relatively short on calcium and phosphorus and hold relatively more potassium.)

The first two editions of my book *Growing Vegetables West of the Cascades* were published in 1980 and 1981. Both editions now seem embarrassingly oversimplified. Re-reading them now makes me realize how entirely wrong a garden writer can be and still help people get a better result. COF had not yet appeared. In 1984, I wrote a third edition, recommending a primitive version of COF. The recipe was as follows (but, please do not use it):

4 parts seedmeal or fishmeal

1 part dolomite lime

1 part rock phosphate or ½ part bonemeal

1 part kelp meal

Blend, spread and dig in 4 to 6 quarts of this mixture over each 100 square feet of garden.

Seedmeal induces rapid growth almost by itself. Over the two to three months it takes seedmeal to decompose, it supplies nitrogen and some phosphorus; the phosphate rock or bonemeal elevates that phosphorus release and gets it into better balance with the amount of nitrogen. Dolomite lime supplies calcium and magnesium, and the kelp brings in any and all needed trace minerals. No potassium fertilizer seemed required in Cascadia. I had been reluctant to put lime into a general purpose use-every-day fertilizer, but I discovered that every acre in Cascadia is leached of 300–500 pounds of elemental calcium every single winter. If I only put in enough dolomite lime to replace that loss, no Cascadian soil should get overdosed, and at least the plants would be ensured of enough calcium and magnesium nutrition. I know now that dolomite is the least ideal kind of lime for most soils, but I discovered that only one year ago.

The fourth edition, entirely rewritten in 1988, stressed that Cascadian soils did not get hot until mid-summer, unlike soils in most of the United States. The rate of nutrient release from the breakdown of organic matter is a function of soil temperature; in chilly soil, the release of nutrients from organic matter breakdown is insufficient to feed demanding crops. Cascadians need to use strong stuff to succeed handsomely. The fourth edition advised separating fertility building from humus building — that it is best for vegetables to get most of the nutrients they need from organic concentrates, much as commercial growers use chemical fertilizers. I suggested that ordinary compost should not be considered a meaningful source of most plant nutrients but as a necessary part of maintaining a healthy soil ecosystem. COF was the centerpiece of my system. I had become unshakably confident about COF; my method was brilliant! Not only did COF have a track record of growing great gardens, I thought its use gradually pushed Cascadian soils in the direction of what William Albrecht considered an ideal mineral balance. Much as J.I. Rodale opined about the perfection of his organic method after only a few years of hands-on experience at Emmaus, I came to think I had figured out all the important stuff in only 15 years. No Cascadian gardeners reported back that COF had let them down. I got many letters thanking me, saying that before COF, poor garden; after COF, brilliant.

It couldn't have been any other way in Cascadia, a cool climate where most homesteaders practiced compost gardening in the tradition of Rodale's *Organic Gardening* magazine. Cascadia is logging country; there's lots of sawdust available free, or nearly free, for the hauling. One way or another, a lot of that sawdust ends up in gardens. Consequently, the typical garden compost was far lower in nitrogen and phosphorus than it might have been and a lot higher in potassium and decomposition-resistant carbons than it should have been. Consequently, my neighbors' crops grew so slowly in chilly spring soils that mustard and spinach put up seed stalks before they had grown large enough to harvest. The pea vines would start out so slowly in that cold nitrogen-deficient soil that they'd be dying from heat-induced virus diseases before they formed peas. Spring cauliflower? Forget that! Heat-loving vegetables started out haltingly in late spring because Cascadia's brisk nights and moderate daytime temperatures had not yet warmed the soil sufficiently. Compounding the problem, Cascadia's frost-free season could be short; even in mid-summer, the nights inevitably ranged from cool to chilly. Consequently, Cascadian corn takes at least one month longer to ripen than it does east of the mountains; most tomatoes are still green on the vine when the first frosts arrive. Cascadians who simply spread some COF and hoed it in got an enormous growth response early in the season. Consequently, they enjoyed a much larger harvest over all, as well as nudging their soil into a better balance, giving their food better flavor and nutrient-density.

How could I have missed?

In 2005, after eight years living in Tasmania, I wrote *Gardening When It Counts*, which was to be sold throughout North America and Australia. This book also depended on COF. Then I really got feedback! The first inkling came when Americans living in semi-arid and desert areas were unsettled by the idea of using ag lime where the evapotranspiration ratio was well under 100. These people had local knowledge, so they asked me if I really meant for them to lime in western Nebraska or in Arizona...and thus it was that I discovered my own ignorance. But tens of thousands of copies of *Gardening When It Counts* had already been sold. An author is never certain that the publisher will issue a revised updated version. So I dithered.

Then my personal life went through a convulsion. Temporarily in rented quarters, sharing my ex-veggie garden with my Ex until I got another garden going, it didn't take long for me to notice a nearby residential subdivision being prepared on 20 quarter-acre lots at the cheap end of an upscale neighborhood. The ditches for laying sewer and stormwater lines revealed a foot of dark brown topsoil and an intensely red-colored, free-draining clay-loam subsoil that, at six feet down, was almost entirely free of rocks. The locals have a name for this beautiful stuff — "red soil," or red krasnozem soil. The term *krasnozem* comes from an antiquated soil classification system; it refers to soils forming from deposits of wind-blown dust where the evapotranspiration ratio is below 100 — like in Kansas or the Ukraine. My red soil here in Tasmania happens to be volcanic, developed out of a type of basalt that decomposes rapidly, making a deep soil that does not form a distinct clay subsoil, as is the usual case where the evapotranspiration ratio is over 100. Instead, Tasmanian red soil becomes gradually more clayey as you dig deeper, but even at six feet deep, it is still open and free-draining, holding a lot of air to facilitate root penetration — in other words, it is a highly productive agricultural resource.

Serendipitous! Class I agricultural soil in quarter-acre lots. Just what a single guy approaching age 65 with my interests wants. So I bought one and built a modest passive solar home intended to demonstrate urban farming at its best. Got my eighth-acre food garden going. And then Annie, who had been one of my students, emerged from the death of her husband three years previously and noticed that this male bird had build a beautiful but half-empty nest. Soon, we married. Annie is as much a gardener as I am, but wanted her own playpen. She had sold her house to live with me and had a pocket full of cash. There just happened to be another red-soil quarter-acre lot adjoining mine still unsold; so we bought it. That lot became my garden; the open sunny eighth-acre on the original house lot became Annie's.

Maybe it was having this big, empty rectangle of new possibility; maybe Serendipity, the universal force, had gotten involved. In 2002, I had self-published a local how-to entitled, not surprisingly, *Growing Vegetables South of Australia*. So I decided my new garden would demonstrate exactly what my local book advised. In other words, I grew this garden precisely by my own book.

Potentially, this quarter-acre lot could be fertile, productive soil, but it was not yet so. The locals told me that before the subdivision went in, the area had been a hayfield — so I knew there had been continual removals. Then it was used as a golf driving range (I still dig up the occasional ball). From at least 1998, when I started living nearby, I had observed the field being mowed (no removals) once a year. It otherwise went unused until it was subdivided. The grass it produced was short and thin. So the first thing on my agenda was to get that quarter-acre enlivened. I brought in 40 cubic yards of poppy marc, an organic industrial waste product similar to mint straw. It looks something like crumbly animal manure, but has little odor. Poppy marc contains plant nutrients, but I didn't buy it for that reason. I considered it worm and microecology food.

So the marc was spread, the field tilled, and the lot divided into 100 square foot, slightly-raised beds, with perennial crops going in around the fringes. From that point on, the only nutrients going into that soil were in COF. By then I had developed a new — and what I considered improved — recipe. Thousands of Tasmanians had been getting great results from it (but please do not use it, there's a better recipe to come):

4 parts seedmeal
½ part ag lime
½ part dolomite lime
½ part gypsum
½ to 1 part phosphate rock
½ to 1 part kelp meal.

The main change was that the ratio of calcium to magnesium was adjusted to further boost calcium, and it now provided some sulfur (gypsum being calcium sulfate). I made that change that because I discovered many soils on Tassie were short sulfur. This recipe also went into *Gardening When It Counts*.

From 2007 until 2011, I used nothing but this recipe on that quarter-acre garden. Each bed got the same treatment: COF at 1 gallon per 100 square feet before every crop, and once a year, whenever it proved convenient, every square yard had one bag of mushroom compost dug into it. The garden's own waste was also composted and returned to the garden. Most beds got COF twice a year; occasionally I squeezed three crops (and three COF doses) into a growing year. I applied another dose of COF prior to planting every crop, *and* I used COF for side-dressing crops that seemed to need a bit of a boost.

During those years, the garden generally grew great, although, after the first year, I did develop a few niggling disease problems. This didn't surprise or much worry me; I'd never had a garden without some difficulties. I failed to connect these troubles with possible soil infertility — not with all that COF going in! My overwintering garlic often succumbed to a type of cold/wet-soil-related root rot, and I had to contend with destructive leaf diseases on onion crops. But then, so did the many onion farmers in my part of Tasmania, so I assumed onion disease spores were present everywhere on Tasmania. So I soldiered on.

Since coming to Tassie, I had been producing my own seed for an open-pollinated slow-bolting summer spinach variety that had to be overwintered in order to ripen its seed in time. But most of the overwintering spinach plants died of disease before starting to form seed. I had assumed (incorrectly) that spinach did not overwinter well because of humidity. I had been consoling myself with the thought that I was making selections for disease resistance. One other difficulty: red krasnozem soil revealed a distressing tendency to pack tight during the winter rains, leading to garlic root rot and making it difficult to work up a fine seedbed in spring if the moisture conditions were not close to ideal when I dug beds. One smart local gardener had solved that problem by mixing large quantities of beach sand into their red soil. I tried that on a small area; it worked, but I wasn't eager to cover a quarter-acre lot with a 2-inch-thick layer of beach sand; doing that would have cost $4,000 at least. Most locals reduced compaction by importing large quantities of organic matter or by maintaining a permanent mulch. I prefer not to mulch in a maritime climate. In spring, mulch slows soil warm-up; over the winter, mulch breeds plagues of primary decomposers (snails, slugs, earwigs, wood lice). I cannot resist the temptation to mention something here that may seem unconnected until you read further: Tasmanian garden centers do not routinely sell ordinary ag lime because we have a notable and highly popular Aussie television garden personality living in Tasmania who strongly extols the virtues of dolomite. The gardener who invented the use of beach sand was lightening a soil that had been tightened up because it had been given too much dolomite.

As I mentioned in the Introduction, I moderate a yahoo-hosted Internet chat group called "soilandhealth." About 2010, soil analyst/author Michael Astera joined the forum. One day, I mentioned that my soil seemed harder than it should considering the high level of organic matter. Michael replied that I probably had put too much magnesium into it — excess magnesium makes clay become sticky and tightly packed. He suggested I try removing the dolomite lime from my COF and see what happened in a year or so.

So, I eliminated dolomite and correspondingly increased ag lime and gypsum so as to not reduce the total amount of calcium in the mixture. A year later, with two or three doses of reformulated COF, the soil was much looser. Michael then offered to give me a free soil analysis. So I had a soil test done and got a major shock. After four years of managing my garden strictly according to my own book, the soil was way off target. I was quite short on potassium, phosphorus, sulfur, zinc and copper. What was worse, I was getting way too high in calcium (according to Michael Astera's system), and my magnesium level had not yet declined enough. Because of that excess calcium and magnesium, my soil pH had risen to 7.0. On the plus side, I had developed slightly over 10% organic matter. There was plenty of boron and manganese, and so much iron that I could have sent my soil to the smelter. No wonder it had such an intense red color. Michael worked out a custom fertilizer to improve my situation. It consisted of precisely weighed and uniformly distributed amounts of:

Canolaseed meal
Monoammonium phosphate
Potassium sulfate
Sea salt
Zinc sulfate
Copper sulfate
Kelp meal
Elemental sulfur.

Michael suggested MAP (monoammonium phosphate) hesitantly, since so many organically inclined people resist synthetic fertilizers. But I did have high pH and calcium in excess of Michael's target level, and that's the sort of circumstance in which MAP shines. I had read Donald Hopkins's book *Chemicals, Humus and the Soil* long before, so I had no problem accepting one of the harmless artificials. No further organic matter was called for, at least not that year, not with 10% already present. So I temporarily gave up buying mushroom compost. I spread only my own compost made from my garden's own organic wastes, and I only used that compost on crops with the most delicate root systems or the highest nutritional needs.

I started using Michael's prescription about mid-spring. Some beds were planted with overwintering crops; on these, I merely scattered the fertilizers between the plants and let spring rains wash them in. The spinach seed crop immediately stopped showing signs of disease; no

Chemicals, Humus and the Soil

Monoammonium phosphate is an artificial fertilizer not approved for use in certified organic production. But any source of phosphorus that is approved for organic agriculture brings with it a good deal of calcium and performs poorly where there is no soil acidity to release the phosphorus. So, when soils are overdosed with calcium or otherwise have a pH over 7.0, I chose MAP over any organic fertilizer. I would never use DAP, which is di-ammonium phosphate. In another chapter, I will discuss MAP and explain more fully why it is a useful fertilizer.

The early organic movement made a primary distinction between natural and artificial fertilizers. The original case conjured up against artificials constantly reappears in books and magazine articles. In those early years, the distinction came with much hostility and name-calling. In that climate, a brilliant guy, Donald Hopkins, wrote a peace-making book called Chemicals, Humus and the Soil. Hopkins explained that if soil is allowed to lose its organic matter content, it declines into infertility no matter how much fertilizer is put into it. But when soil organic matter levels are maintained and the lime used up by fertilizers is replaced, nutrients are nutrients. Pretty much.

Hopkins' book is available for free download at soilandhealth.org.

more plants died, and the survivors began growing rapidly. The garlic crop started growing faster than in previous years. The usual Allium leaf diseases didn't appear. When it came time to dig that garlic, the heads were half again larger than before, and there was no sign of root rot. And when I harvested the spinach seed, I never saw such a yield. I got about ten pounds of fat, strong seed from about 150 square feet of bed — pretty good, I'd say.

The remineralized spring veggies tasted better than ever before — richer, more complete somehow. The spinach we steamed that spring was so sugary, I almost didn't like the flavor. Annie has several regulars who take home a mixed box of vegetables once a week. Our customers said our vegetables that year tasted much better than they had the year before — and the year before they had been telling us that our vegetables were the best they'd ever tasted. But that was before I remineralized the soil.

Better-mineralized vegetables also gently improved our health a few notches, with one slight drawback: Annie and I started gaining weight even though we were consuming mostly vegetables; we hadn't yet learned to adjust our intake down to match the increase in how much we were eating because the vegetables all tasted so great!

After remineralizing, we had even less interest in buying treats, meats, cheese and other things from the supermarket. In other words, homegrown veggies became a larger fraction of our total intake than they already had been. This shift was effortless; we're eating what we enjoy most. I think improved vegetable nutrition has enabled me to mostly give up our excellent Tasmanian cheeses (much of the time), consequently, I am not having as much discomfort at night. I also have more energy — important when a bloke gets to age 70.

The English language has few words to accurately describe flavor. But how about this attempt: we have long enjoyed eating zucchini splosh. To make splosh, you steam or simmer chunks of zucchini until they are soft enough to mash. Then you mash. While mashing, add a big pat of butter and a little black pepper. Salt if you must. That's it. This year, our splosh tastes nearly as rich as a savory pumpkin soup. It's incredible. We *want* to eat a big bowl of zucchini splosh every night. We were sad last autumn when we ate the last serving the garden would provide until the next summer.

And our sweet corn! I hadn't tasted corn that good since coming to Australia. I'd been complaining about the lack of good-flavored sweet corn varieties in Australia. I discovered that one reason was a quarantine restriction on corn seed imports. The main result of this restriction has been to create a protected market in which our domestic seed producers can charge several times the price Americans have to pay; to add insult to this injury, we home gardeners are offered only a handful of second- or third-rate varieties. While in Australia, I've done trials that included every corn variety legally available, but remembering the corn trials I did when I had Territorial Seed Company, I would say that in Australia I have never tasted a variety I would have scored over 7.5 out of 10 because I still remember the flavor of Jubilee, or Sugar Dots, which I generally awarded a 9.5. After remineralization, a variety I scored 7.0 last year tastes like an 8.0 this year. And I'm expecting 8.5 next summer as more nutrients leach into the subsoil. Remineralized soil!

Which mineral on that list made the difference? I do not know; and frankly, it doesn't matter. There are no unimportant plant nutrients. Elements used in only tiny amounts, like copper or zinc, can have effects as major as nutrients used in large quantities, like phosphorus and potassium.

Using COF Now

If your intention is to produce nutrient-dense food on a scale that means a great deal to the family economy, do a soil test, and amend the soil in the direction that maximizes nutritional outcomes. That's the best way. Thinking just in terms of money, if you're growing a large-enough garden that its output makes a financial difference, and if its fertilization requires the purchase of anything at all, why not add another $20 to your annual cost and do a soil test first? Then you can buy only what the garden really needs. The test could save you more than its cost. And if you think of it in terms of your family's health, there is no choice at all.

But if yours is a small garden that doesn't seem to justify the cost or effort, if your food garden is not a discrete area but just a few vegetables interplanted amongst flowers and other ornamentals, or if it is in small, irregularly sized beds, each with highly different natures, soils, histories of being amended, and so forth, or for whatever reason being guided by a soil test seems undoable, then your problem can be solved by fertilizing with a fairly complete and balanced organic fertilizer recipe.

The major concern when designing a COF is achieving as much *balance* as possible without creating excesses. Deficiencies are easy to remedy; excesses...well, as Hugh Lovel once joked, "it's easy enough to resolve soil nutrient excesses, no more difficult than getting too much salt out of the soup." My COF is designed to, above all, avoid creating excess; therefore, it cannot completely ensure there are no minor nutrient or trace element deficiencies. There is no way out of this problem except to custom-design a new COF every year or two from soil test results.

Making COF yourself requires that you first obtain up to ten ingredients. (To source them all might take a bit of clever shopping because garden center merchants as yet don't expect home gardeners to request some of these substances. I hope that will change.) Making COF will involve nearly the same effort and expense as would biting the bullet, getting a soil test, and formulating something perfectly suited to your land. And no COF can possibly grow food to the degree of nutrient-density that can be achieved from remineralization according to a soil test result.

Making COF requires measuring ingredients by volume using ordinary kitchen gear and then thoroughly blending and uniformly distributing the material. I measure out fertilizers with a quart-sized worn out Teflon-coated saucepan and a cheap, plastic half-quart measuring cup. For trace elements, I measure rather more accurately, using a kitchen measuring spoon set. Perfect accuracy is not required; plus or minus ten percent is good enough. I am certain as you read the recipe, you'll have questions or may not know what some of these substances are. Don't worry, I'll give you a lot more information about these materials in later chapters.

Complete Organic Fertilizer

To make enough COF to generously cover 100 square feet, mix:

3 quarts oilseedmeal such as soybean meal, cottonseed meal or canolaseed meal

or else:

1½ quarts feathermeal or fishmeal (smelly)

or, the very best combination is probably:

2 quarts oilseedmeal, 1 pint feathermeal and 1 pint fishmeal

plus

1 quart soft or colloidal rock phosphate (the best choice by far), or bonemeal

1 quart kelp meal and/or 1 pint Azomite (for trace minerals)

And/or apply liquid kelp every 2 weeks as a foliar throughout the season.

Lime: choose one of these two options:

If you garden where the land originally grew a forest, add these two:

1 pint agricultural limestone, 100# (fine grind) and

1 pint agricultural gypsum;

or else:

If you garden where the land originally grew prairie grass or is a desert add:

1 quart agricultural gypsum.

If you do not live in Cascadia, add 1/3 cup potassium sulfate.

You may consider the following last four items optional:

1 teaspoon laundry borax or a smaller quantity of Solubor (½ gm actual boron)

1½ teaspoons zinc sulfate

2 teaspoons manganese sulfate

1 teaspoon copper sulfate.

When all ingredients are in the bucket, mix them very thoroughly before spreading. I use either of two mixing methods: 1) Slowly pour the materials from one bucket to the next and then back. Repeat this about six times. Or, 2) stir the contents with my hand. The first method works the best, but can raise a bit of dust and is best done outdoors.

Gardening with COF

When preparing an already fertile bed for planting, first spread compost ¼ inch thick. That's a thin scattering, but if done once a year, it's enough to maintain an existing high level of organic matter and even increase it slowly, if you're using high-quality truly mature "humified" compost of the sort I encourage in Chapter 9. You can also use well-rotted manure or incompletely ripened compost spread twice as thick. Over the organic matter, uniformly spread COF at the rate of 4 quarts per 100 square feet. And then dig it all in. If you garden "no-dig," then simply spread your soil amendments and then mix them in shallowly with a rake or hoe, if and where you can. If the garden is arranged in traditional long rows, then the place for the COF (and compost) is a broad band about 2 feet wide, with the seeds or seedlings placed down the center of this fertile strip. For this arrangement, I suggest spreading 4 quarts COF over each 50 row feet. After digging, it's best (but not absolutely necessary) to delay sowing for a few weeks, to allow the nutrients to blend into the soil and its ecology as well as letting the soil settle. This will restore capillarity, a natural movement of moisture toward surface, helping to keep germinating seeds moist when you do plant them.

The amount of COF to spread may be adjusted to suit the soil's capacity to hold plant nutrients. The next chapter will explain this in terms of "light" and "heavy" soils. Generally, sandy soils are light; clayey soils usually are heavy — unless you live in the southeastern United States, where the clays are old, tired and *very light*. Light soils respond powerfully to COF at 4 quarts per 100 square feet. However, heavy soils may need more than 4 quarts the first few times you use it — 6 quarts is a better amount for them. After a heavy soil has been fertilized for a few years, it should not need larger amounts of COF than a light soil does, and likely will come to need less ongoing refertilization than a light soil.

After you have distributed COF a few times, you'll learn to gauge by eye how thick it needs to be spread in order to make the plants grow fast; gauging it becomes intuitive. Most gardeners using COF for the first time are surprised that their plants can grow so rapidly. Once you see this happen, if you later err by applying too little, the crop won't grow as fast as you know it can, and you'll then side-dress more. If too much COF goes in and the plants grow excessively fast (if that is even possible), well, no damage done, and you can brag about that 3 pound broccoli the diameter of a dinner plate.

If plant growth slows during the crop cycle or if growth did not seem rapid enough from the beginning, then side-dress with additional COF at half the starting rate — 2 quarts per 100 square feet. (Side-dressing means sprinkling the fertilizer over the ground that the plants *will* be growing into in the next three weeks and then shallowly hoeing it in, if possible.) If a side-dressing produces no growth response, it was not needed and should not be repeated.

If you are refertilizing for a second crop in the same year, spread and work in a half-dose of COF, no more compost should be required. Two quarts per 100 square feet should do. Low-demand crops following ones previously given COF probably do not need another application. Note that if you're going to depend on COF, there's information you need in the next chapter, where I discuss how to adjust the amount of nitrogen fed to soil according to the needs of the crop. You can adjust the amount of COF in this same manner.

What I've just given you is a complete, workable soil-fertility maintenance system that will produce nutrient-dense food — a small amount of compost once a year, a layer only ¼-inch thick, and a dose or two of COF. You'll gradually develop plenty of soil organic matter; you'll have plenty of minerals in the sort of balance that makes plants grow big, fast, healthy and with more nutrition. I know a ¼-inch-thick layer of compost seems insufficient to most gardeners. But unless you've the misfortune of trying to grow vegetables in dense clay, building extremely high levels of organic matter is the least desirable way to lighten soil. A far more effective way to loosen up soil is by balancing the calcium-to-magnesium ratio, and my version of COF does that for you automatically.

COF For Sale

If your garden is small, it may not seem sensible to buy individual ingredients in 50 pound bags. In that case, purchasing something ready-made is sensible. However, the per-pound cost will be higher when you buy it premixed. And you're going to get someone else's idea of what constitutes good balance.

I live on a remote temperate South Pacific island. Despite that financial disadvantage, making my own COF from locally sourced ingredients obtained in farm-sized sacks costs me about 50 cents per pound. Pre-mixed organic fertilizers inevitably cost more. To help you make a well-informed choice, a quart of COF weighs about 2 pounds. To fertilize 100 square feet takes about 8 pounds.

Complete *inorganic* fertilizers have been around ever since Victor Tiedjens developed the art of hydroponics. Miracle Grow, Peters and similar products are their current incarnation. In the late 1970s, Concentrates, an agricultural distributor in Portland, Oregon, manufactured a COF sold in 50 pound bags. I never used it then because the price per pound seemed steep to me compared to mixing up something myself. As I learned more about the region and handling its soils, I realized that their COF formula was incorrectly balanced for Cascadia — the label said 5-5-5, and I wanted something more like 5-5-1. Concentrates must have learned a few things over the years; now their home-brand blended organic fertilizer is 5-5-3, and it costs less than $1 per pound. It looks like the best dollar value for someone close in that region. Concentrates has low prices, in general.

My books have been steady sellers in Cascadia since 1980. They created a demand for COF. Around 1982–83 a wholesale distributor of garden-related supplies, Down To Earth (DTE), in Eugene, Oregon, began manufacturing a range of complete organic fertilizers. Someone doing this commercially has advantages over the home gardener; they can include ingredients only obtainable by the ton in bulk, so they can build a more complex mixture that releases some nutrients rapidly and others gradually, over a longer period. DTE's fertilizer blends are approved for organics. Down To Earth can't ship their products to all states because each state requires registration of all (mixed) fertilizers before they can be sold there. But DTE does service the entire Pacific Northwest.

Black Lake Organic, in Olympia, Washington, makes a range of superior complete organic fertilizers with a range of NPK analyses they call "Bloom." Fifty pound sacks cost about $1.75 a pound. Black Lake's flagship Bloom No. 1 fertilizer 4-5-3 was designed by owner Gary Klein. Gary is an admirable guy who really cares about people; I've known him for a couple of decades and am pleased to have this chance to support him. Gary's fertilizer is as complete and as cleverly balanced as it is possible to make a pre-mix. Black Lake Organics also sells a full range of OMRI-approved fertilizers by the 50 pound sack and — even better — they will weigh out individual fertilizers for Internet customers by the pound. No matter where you live in the United States, if you only need one pound of zinc, copper or manganese sulfate, Black Lake Organics will ship it to you at a reasonable price.

Food gardening is a counter-cyclical activity. When economic times are good, people choose to take summer vacations. When times are hard, people grow food gardens. Right now, interest in home-food gardening is undergoing a major resurgence. So, new blended organic fertilizers are coming on offer throughout North America as new businesses make OMRI-listed soil amendments and natural concentrates (like seedmeal) available. Shipping costs and state registration requirements tend to limit how far these can travel, and the situation makes it foolish for me to attempt to provide a complete source directory or reviews of prepared fertilizers. However, since I have so easily discovered so many versions of COF on offer along the northern half of the west coast of the United States (the part of the United States I am most familiar with) then your area is likely to provide some options as well. You'll find a few leads in the Appendix.

Tweaks

The soil's capacity to hold calcium, that is, its need for lime, varies. The amount of calcium in COF can be adjusted correspondingly. The recipe I provided a few pages back is designed for light soil one with little clay content. If that recipe were to be used on a heavy soil it could take many years to build up the level of calcium needed. If you garden a heavy soil, I suggest you double the amounts of agricultural lime and gypsum for the first two years and then revert to the basic recipe. Make no other changes. If you have any doubts about your soil being light or heavy, assume it is light; when fertilizing, it is always best to err on the side of less.

I have mentioned that magnesium makes soil get tighter. There's a good deal more about this topic in the next chapter. In the event you garden on a coarse, sandy soil that is so loose and so open that it won't hold water (so loose that your carrots are almost pulled out of the ground by strong winds), you can gradually improve that situation by changing the type of lime you put in COF from ordinary high-calcium agricultural lime to dolomite lime.

If you live in the southeast United States, even if you garden a clay soil, assume it is a light soil.

Why COF Works So Well

COF contains a lot of calcium. It comes from three sources: agricultural lime, gypsum and soft rock phosphate. Although lime and gypsum are the most inexpensive of fertilizers, they may be the most important ones. A 4-quart-per-100-square-foot standard application of COF spreads about 700 pounds of calcium to the acre, a bit short of what one ton of ag lime (40% calcium) would bring with it. Spread that much lime once a year and what the usual agronomist would consider the lime requirement of almost any soil will have been met after a few years. So is it possible to bring too much calcium to the party if you repeatedly apply COF? Consider what Victor Tiedjens said about this in *More Food From Soil Science* (his use of the term "calcium requirement" in this case means the entire capacity of the soil to hold cations):

> *Imagine clay and humus being a series of shelves made of iron and aluminum, and the stuff on the shelves to be the ions, such as calcium, magnesium, potassium, sodium, manganese, and so on. The shelves are deep and the ions on the front may be obtained more readily by the roots than those on the back. Now, imagine the root of a plant being a truck that backs up to the shelf to load up. It needs certain ions. If it gets what it needs freely, the plant grows normally. But suppose those shelves are loaded with potassium and nothing else. Then the plant doesn't get calcium and magnesium. It gets too much potassium and stops growing. But suppose the shelves are almost empty and only hydrogen ions are present. They are gaseous, and the plant can't grow by taking in gas. In addition, the bench begins to deteriorate [due to high acidity] and the root takes in parts of the shelf — iron and aluminum. The root shrivels and dies. It is poisoned. In other words, we must keep those shelves strong enough and full of calcium, magnesium and potassium — in the right proportions. ...Calcium is the one most often lacking.*
>
> *...It is necessary that a soil be limed to a pH equal to 85 per cent of its calcium requirement to support best conditions for growth of crops. For instance, if a sandy soil has a calcium requirement equal to one ton of limestone in an acre-foot, it is necessary that 1,700 pounds of limestone be added to bring the top 7-inch layer into good condition. And to improve the soil down to a depth of three feet, we would have to use approximately 7,600 pounds of lime. In an acid [heavy] soil it may be necessary to put on 15 tons of limestone per acre to supply*

> *the necessary calcium to a three-foot depth. Maximum growth may not be obtained until this is done.*
>
> (p. 82)

Trouble is, if 15 tons of limestone are spread at one go, the soil may get a severe case of indigestion for awhile; perhaps for some years. COF, on the other hand, *gradually* limes the soil at one ton per acre. One ton does not disrupt the soil process. This calcium gradually percolates into the subsoil; the gypsum in COF facilitates that penetration. Gradually, an ever-increasing depth of soil gets saturated with calcium and thus, it opens to root penetration, leading to an expanding moisture supply, expanded access to plant nutrients, and far greater plant health.

As the soil becomes saturated with calcium, something miraculous occurs: it gets better at delivering all aspects of plant nutrition. Picture it this way: the soil's deep storage shelves are far easier to unload from their outer edges. The nutrients far back on the shelves are not nearly as accessible to plants as the ones on the edges. A soil test may show them present, but the plants can't make use of these cations. Lime stocks those inner shelves with calcium; Tiedjens' goal was to saturate 85% of that inner shelf space with calcium, leaving only a small space for additional elements on the outer fringes of the shelving. Newly applied fertilizer gets stored on these fringes, where the plants can access it readily. In practice, this brings a huge reduction in fertilizer cost to farmers and a huge increase in the effectiveness of fertilization. Tiedjens found that, once the soil was saturated with calcium, he could grow a huge crop of corn or soybeans using about one-tenth the quantity of fertilizers a typical farmer thought was needed to produce a similar result. And that is why the fertilizer industry made sure you never heard of Victor Tiedjens — lime is cheap; fertilizer is dear.

Unless you've previously limed heavily when you first begin to use COF the soil's nutrient-holding capacity will not yet have been saturated with calcium. The fertility elements you put in locate themselves on the inner parts of the shelving and are not as easily available. And in consequence, it takes 4 to 6 quarts of COF to prompt enough growth. After a few doses of COF have saturated the topsoil's shelving with calcium, it will seem to need less fertilizer, and you will naturally spread a bit less. A few more doses of COF and you'll start saturating the subsoil. Then you'll be spreading only 2 quarts instead of the 4 quarts it took a few years back. Thus, you will correspondingly reduce the amount of lime you are adding as the soil develops less need for it. After you've used COF for four to six years, you'll be spreading much less and getting much more from it. Don't forget, calcium is constantly leached from soils where the evapotranspiration ratio exceeds 100, and a half-ton of lime per acre per year is about what it takes to replace that ongoing steady loss.

Initially, the tiny quantities of copper, zinc, manganese and boron in COF will not make a difference if a soil is critically short one or more of these elements. Fortunately, most soils do not become seriously deficient in trace elements until a few crops have withdrawn their requirement. But after the soil has become well-saturated with calcium, that little bit of zinc or copper you're adding with COF will be enough to feed the current crop without overdosing the soil. The only way to supply trace elements more effectively than this is to do a soil test and add what the test indicates.

Chapter 5

Remineralization

Soil analysis may seem daunting, but if you've gotten this far into the book, I'm sure you're up to it. The process requires you to methodically measure and carefully distribute soil minerals. It also requires some basic arithmetic. Right now, you may thinking gardening is not supposed to be so complicated and you're in the wrong book. Well, maybe you're not. Maybe this book *is* for you. It will show you, simply and step-by-step, how to crunch the numbers and work out a remineralization program for yourself. I encourage you to at least read this chapter once through lightly, without attempting to make it actually work for your garden; you may be surprised by what soil testing can accomplish.

So how "smart" or how well educated do you have to be to master soil analysis? The truth: not very. I wrote this book to function like "Analysis for Dummies." I will tell you only what you absolutely need to know — in the simplest possible terms. I cover only the bare essentials, leaving out all that fascinating (or boring) background information enthusiastic writers usually can't keep to themselves. For me, personally, the study of soil chemistry and the contemplation of what might constitute the ideal soil and how one can create it is a marvelous puzzle that can endlessly occupy my thoughts. It's possible you don't feel the same way about it.

However, I'm pretty certain that when you *taste* the result, you will be inspired to learn more. And that's why I include mentions of some of the other interesting books out there as often as I think I can get away with it. For me, there's always endless heaps of fascination to delve into. But learning that much is entirely unnecessary if all you want to do is successfully produce nutrient-dense food.

I have simplified the process by supplying worksheets, which you will find in the Appendix. On a single sheet of piece of paper, you will have all the key facts and all the crucial numbers; you will not have to memorize anything. All the arithmetic is on the worksheets, so you won't be looking back and forth through this book trying to find some detail. You will have to grasp a little soft science that I will present much like a PC expert explains the functions of a PC to a non-specialist. So if you never had the pleasure, or never passed, or never honestly passed, a course in high school level inorganic chemistry — no matter. What is really important is that I once did.

You won't need chemistry, but you absolutely must be able to do simple arithmetic well. Not advanced math. Just primary school arithmetic — add, subtract, divide, multiply, fractions, decimals, ratios and proportions. A great many Americans were crushed by their schooling. They still strongly dislike arithmetic; they'll refuse to grapple with it. If you're in that group, be comforted: Erica and Alice Reinheimer, garden soil analysts living on the central California coast, have developed an alternative to arithmetic; it's an online web app (you can access it at GrowAbundant.com) that uses the same target levels I suggest. Transfer the results of a standard soil test to their program (line for line, it looks exactly like a Logan Labs report sheet), and voilà! Your soil prescription. An example of the spreadsheet analysis is shown at the end of Chapter 7 in Figure 7.2.

If a personal computer can do the analysis in milliseconds using a program that costs less than ten bucks, why should anyone want to bother to do it manually? There are two excellent reasons: 1) to fully appreciate the web app's answers; and, 2) to allow greater flexibility — it is sometimes cheaper or better if *you* figure out the materials to use and how to combine them. The computer program is extremely useful, but if you really understand soil, you sometimes can make smarter choices than any computer program.

The numbers on a standard soil test report represent what is available in that soil to feed your crops. The arithmetical system this book is based on lets you quickly work out what those numbers should be *ideally* and what soil amendments are needed to bring your soil to those targets. After going through one example soil audit, you'll be equipped to do one for your own garden. I predict that by the middle of your first remineralized growing season, you'll be encouraging your gardening friends, neighbors and family members to get their soil tested — and you will probably be doing the arithmetic for them. You may even become the neighborhood's garden soil analyst. The main reason I am writing this book is that I hope you will.

The Target

An archery target usually consists of concentric rings with a bull's-eye in the center. When balancing soil, the target is the relationships among six elements: calcium, magnesium, potassium, sodium, sulfur and phosphorus. The other plant nutrients — boron, iron, copper, zinc and manganese — are equally important, although they are not added in large quantities, and we are not as certain about where their bull's-eyes are.

There are a handful (or maybe a hatful) of other elements that plants don't seem to absolutely require but do pick up in tiny traces; and there are a few elements plants do absolutely require for their own internal chemistry, but only in the slightest of traces, like molybdenum and cobalt. Rest assured, I don't overlook any of these elements, because even if the plants don't seem to require them, your body does.

The Science

There are about 100 known elements; many of them are familiar, like gold, copper, oxygen and iron. This book is concerned with 11 elements that nourish plants. In the next few paragraphs, you'll find the majority of the chemistry you need to master this subject. I give you the names of the elements important to soil fertility, the chemical symbol or abbreviation for each of these elements (which you will have effortlessly memorized by the time you've read this chapter once through), and key facts about some of the elements. If you suffer from Post Traumatic Schooling Disorder, I say, relax!

When elements dissolve in water, the material separates into individual atoms, called *ions*, that carry a faint electrical charge, either positive or negative depending on the element involved. Ions with positive charges are called *cations* (pronounced "CAT-ion," not "cay-shun"). Ions with negative electrical charges are called *anions* (pronounced "AN-ion"). Ordinary table salt, NaCl, is a combination of one sodium (Na) cation and one chlorine (Cl) anion; when dissolved in water, table salt splits apart into a cation of sodium and an anion of chlorine. When the water evaporates, they recombine because their charges attract each other, like opposite magnetic poles. Once brought back together, they link, and we have NaCl again. There are a great many such combinations, such as zinc sulfate or calcium carbonate; all these combinations are called *salts*. Many salts dissolve in water, sometimes readily, sometimes only barely and reluctantly. A salt carries no electrical charge until it is dissolved.

Table salt, NaCl, could also be written Na^+Cl^-. As I just said, when you dissolve table salt in water, it splits into the cation Na^+ and the anion Cl^-. Evaporate the water, and these charged particles rejoin and reform crystalline table salt. The terms *cation* and *anion* appear frequently in this book; do not allow yourself to be confused about what they mean. If you know next to nothing about chemistry, be comforted: there are only a dozen or so technical words you will have to grasp and remember. To assist you, there is a short Glossary in Chapter 1. I suggest you take a moment to go back to it and read it over so you are familiar with the terms before reading further. And you should familiarize yourself with the names of the elements and their symbols in Table 5.1. But remember: There is no need to memorize anything.

Growing nutrient-dense food requires bringing nutrients in soil to target levels that are in balance with other nutrients, while at the same time making sure there is a healthy soil ecology helping the process along. Creating maximum soil fertility is not necessarily about having *more*; it is about achieving *balance*; often, it is about having *less*. My underlying strategy is to present both the plants and the soil ecology

A furrowslice acre: a six-inch-deep soil layer covering one acre. It weighs about 2,000,000 pounds.

with a luxurious abundance of everything they can use — except for potassium. About potassium, there will be much more, soon.

Sampling

When assaying soil, a sample is soaked in an extractant solution; then, the elements removed from the soil by the extractant are analyzed. This book focuses on a type of soil test that uses a Mehlich 3 (M3) extractant. A standard soil test using the M3 method accurately measures the available quantity of 11 essential plant nutrients. To adjust for differences from spot to spot in any field or garden, several samples are thoroughly blended before the extraction is done. The test result will be accurate only if each soil sample going into the blend is the same size. For gardening purposes, we usually analyze the top six inches of topsoil because that is where most of the biological activity happens. It's where the crop does the majority of its feeding, and it's also where we can conveniently mix in fertilizer with a shovel, fork or tiller (or plow, spading machine or disc harrow).

Element or Compound	Symbol
The Cations	
Calcium	Ca^{++}
Copper	Cu^{++}
Iron	Fe^{+++}
Magnesium	Mg^{+}
Manganese	Mn^{++}
Potassium	K^{+}
Sodium	Na^{+}
Zinc	Zn^{++}
Ammonium	NH_4^{+}
The Anions	
Boron	B
Carbonate	CO_3^{-}
Chlorine	Cl^{-}
Nitrate	NO_3^{-}
Nitrogen	N
Oxygen	O
Phosphorus	P
Phosphate	PO_4^{-3}
Sulphate	SO_4^{-2}
Sulfur	S
Trace Elements	
Chrome	Cr^{+}
Cobalt	Co^{+}
Iodine	I^{-}
Molybdenum	Mo^{-}
Selenium	Se^{-}
Vanadium	V^{+}

Table 5.1: *Chemical Symbols of Important Elements.*

Busting open compacted *subsoil*, assaying it separately, and remineralizing it may prove highly rewarding — if you have the endurance or the machinery with which to accomplish that task. However, mixing topsoil with subsoil is counterproductive. And mixing fertilizers and/or compost into home-garden subsoil requires laborious — and generally unnecessary — double-digging. So, six-inch deep samples it is.

Soil varies from spot to spot, so you need to take several samples and blend them to determine average values. An established home garden will have different fertility profiles from bed to bed because preceding crops will have removed differing amounts of minerals. Plus, few gardeners feed every part of their garden the same way every year. Even backyard lawn soil can vary from place to place. So, if your garden is tiny (100 or 200 square feet), take four or five samples; if you're sampling an area that's 1,000 square feet (the size of the average backyard garden) to 10,000 square feet (a quarter-acre), take eight to ten samples. Each sample must be a uniformly thick slice from the surface to the maximum sampling depth (usually, six inches). Professional consultants use a soil probe that quickly samples soil to a precise, repeatable depth; the probe is a thin-walled stainless steel pipe that is pushed or augured into the earth. Each sample it takes is a cigar-sized cylinder of soil six inches long. If you were sampling soil for a living, you'd want one of these tools.

Give a moment's thought to how you will end up with a reasonable average of your soil. There will probably be places in your garden to avoid taking samples from because the soil is for some reason quite different from the rest; including samples from these spots will incline your result away from what will most benefit the largest area. Avoid sampling within five feet of building foundations; during construction, concrete, stucco, mortar, paint, sealants, etc., are often splattered on the ground, then covered by just enough topsoil to level the grade. In fact, you should avoid growing food crops close to buildings because of the high risk of contamination. Avoid sampling spots where you have spilled fertilizer or chemicals — even laundry detergent. If it is a pasture you're sampling, avoid spots where the grass grows particularly lushly, because it's probable some beast urinated there. Avoid sampling low, wet spots unless the whole garden is a low, wet spot. Don't sample sides of embankments, places where hay has been stacked, or places where there was once a chicken coop (unless that area is most or all of what will be your garden). Neal Kinsey, a highly regarded soil analyst and author, recommends not sampling where buildings once stood; he was probably thinking of farms, but there are now many urban situations where this may apply. If there is a minor part of the garden that grows things much better or much worse than the remainder, and you are not willing to do a separate soil test especially for that area and fertilize that area differently from the rest, do not sample there even though you plan to spread fertilizers there. Your most efficient use of soil testing is to bring the largest possible part of your garden into balance; let the odd bits come along as best they can.

Take samples with a spade or garden trowel that you first scrub clean; it's best to use stainless steel because any flakes of rust (iron) or loose galvanizing (zinc) getting into the sample can hugely elevate the reported iron or zinc content. Strip aside any mulch or loose organic matter before making your hole. Be aware that the first half inch of humusy soil immediately below mulch can be much richer than the rest, but it must be included in the sample.

Once you have your samples, mix them together. Erica Reinheimer puts all her samples into a clean glass kitchen mixing bowl — a *very* clean one. I use a food-grade plastic bucket that never had any fertilizer, detergent or other chemicals in it.

When all the samples are in the bucket or bowl, mix them thoroughly with a clean stainless steel spoon. Thoroughly! Remove any pebbles or bits of vegetation (or worms). There is no need to thoroughly dry the sample, although doing so will lighten it, saving you a bit of postage. When it gets to the lab, they will grind it and get it properly, accurately, scientifically dry.

Soil labs require something like half a pound of soil; lab websites provide the details, as well as more information about taking samples. Logan Labs, for example, wants a full teacup — unless you have already air-dried and sieved the sample so it is free of stones, worms, bits of organic matter, etc, in which case three ounces will do.

To measure fertilizer, you're going to need an accurate kitchen scale; the best type is an electronic scale that can handle up to 10 pounds. I urge you to get one that also displays metric measurements and is accurate to one gram. Some fertilizers will be used by the pound or half pound per 100 square feet; others are called for in tiny amounts that are far better measured in grams rather than tiny fractions of an ounce.

You may as well buy that scale now and use it to weigh your sample. Put your soil sample into a *new* Ziploc bag. Mark your name and the sample details (if you're sending in more than one) with an indelible, permanent marker on a strip of paper taped to the *outside* of the bag (inks can contaminate the sample). If you're sending in more than one sample, make sure the labeling is clear and the labels are securely attached. I've seen "indelible" markings on Ziploc bags mysteriously fade away. Be sure to fill in the soil lab's transmittal form.

Soil testing does not cost a great deal in North America. As of 2012, you can get an M3 test for 20 bucks or less. Paying more money for an M3 test doesn't necessarily get you more value; it may not even get you more personal attention. And once you've read and learned what's in this chapter, you'll not need an analyst's opinion about what to do.

Soil testing methods are named after the extraction method used. Just so you don't get surprised by some never-before-seen term, the most commonly mentioned extractants are: Bray, Olsen, AA (ammonium acetate), Morgan, Paste Test and Mehlich 3. Some of these extractants are mild chemicals, so their results reflect only what is already dissolved in the soil water. The paste test uses distilled water as the extractant, which truly images the soil water. The Mehlich 3 extractant is a complex acid — about as strong as household white vinegar. Soil labs also do tissue analysis to find out what a plant is actually managing to uptake, compared to the levels the soil test reveals. Farmers use tissue analysis and paste testing to figure out what to do to goose a slow-growing crop into performing again.

Should you attempt to fit the reported levels given by any extraction method other than M3 into this book's system, well, the numbers simply won't work. Later on, after your garden has been remineralized, you may want to know more about soil testing and the whole science behind the minor miracle it produces. I hope you will be enticed to do some serious reading. When you do, you will encounter the other extractants mentioned above.

While considering sampling, your thoughts may have turned to spreading fertilizers. You're going to be feeding the soil with concentrated nutrition that plants respond powerfully to. All fertility elements should be spread uniformly and at the correct rate. If you should accidentally double-apply them when the first dose was already at the highest safe application level, you may well push some substance to toxic levels. This is especially true for boron and copper. It is easy enough for an experienced farmer to accidentally double-dose their soil. All they have to do is overlap the fertilizer spreader on one pass down the field. For that reason, even though the soil analysis may show a large deficiency in one element, for the sake of safety, this book does not risk heavy applications all in one go. The amounts I recommend in this book will grow great crops; and there'll be plenty of occasion next year or next crop to build levels further.

Often, the quantities needed to balance a soil are far greater than the quantity needed to grow a great crop. To protect you from overdosing your soil, this book suggests application limits that will more than adequately grow the current crop and build background levels so that next year's soil audit will show a higher level than this year's did.

Fertilizer application rates will be specified as so much weight of material per so much area. Sometimes it will be only a few grams per 100 square feet. Occasionally, it'll be a pound or two per 100 square feet. Considering the mathematics involved, it works out that the most convenient measurement for the gardener to use is grams per 100 square feet. Then, if an area is 1,000 square feet, you can simply multiply the weight by ten. Or, if your bed is 75 square feet, multiply the quantity needed for 100 square feet by 0.75. Grams can be used because there is a rough equivalence — not exactly equal, but close enough — between pounds per acre and grams per 100 square feet. If the analysis calls for say, 250 pounds per acre, you get that by spreading 250 grams per 100 square feet (or 2,500 grams per 1,000 square feet).

If your food garden is a happenstance hotchpotch of irregular spots set in and amongst ornamental beds, I suggest that you take a considerate walk around your place and have a good think about how much area each growing area is occupying. Consider measuring them all and making a rough map showing bed sizes. If all of this weighing and measuring seems like too much trouble I do understand; but I can't sympathize. Someone who truly appreciates how essential it is to eat nutrient-dense food would never settle for using just the odd bits of a garden to only grow just a bit of food — not in a world where highly nutritious foods can't be reliably purchased at any price. In my opinion, the family food garden should be as large as possible (up to the point that it produces 75% of everything you eat) while juggling the demands of children at play and partners growing flowers.

A large garden is far easier to manage when the space is divided into beds of roughly the same size. Mine are 4 feet wide by 25 feet long, so 100 square feet. That's a generous size bed for the home garden. One such bed can produce all the carrots, beet root and parsnips our kitchen needs for an entire winter. And then, in spring, the whole bed can be remineralized, given a dose of compost, and be sown with a different group of crops. One hundred square feet yields enough Brussels sprouts to keep us and a few neighbors satisfied through the winter. Another such bed amply holds 16 autumn/winter-heading cauliflowers of assorted varieties, all sown at one time but harvested over three chilly months. An adjoining pair of beds, combined by temporarily digging up the path between them, makes a 10-foot by 25-foot winter squash (pumpkin) patch or corn patch. One bed holds all the zucchini and cucumbers we and our neighbors could possibly use.

When all the crops in one bed are sown at the same time and need roughly the same duration in the ground, you can prepare the whole bed for replanting at one time. That's the key to conveniently and accurately remineralizing a garden *and* working the garden efficiently. Plan so you can prepare an entire bed all at once. Otherwise, no matter how good your memory, you run the risk of double applications of fertilizer and skipping spots.

In order to conveniently grow really nutrient-dense food, you may have to reorganize your garden. That's probably a good thing. Every time I build a new garden, it has been better than the previous one.

Soil Labs

UNITED STATES

Here's a pair of excellent soil labs that use M3 extractions:

- Logan Labs, 620 N Main St, PO Box 326, Lakeview, OH 43331, (888) 494-SOIL, loganlabs.com, email: susan@loganlabs.com. Their "standard soil test" cost $20 in 2012. Logan's reports are easy for amateurs to use. I suggest that all newbies to this art use Logan.

- Spectrum Analytic, 1087 Jamison Rd NW, Washington Court House, OH 43160-8748, (800) 321-1562 or (740) 335-1562, spectrumanalytic.com. You want their "S3 test." It costs less than $20. Spectrum is a professional's lab; because their reporting system is complex, I do not recommend them for amateurs. However, I have an account with Spectrum.

USING US SOIL LABS FROM OUTSIDE THE UNITED STATES

American soil testing lab prices are low and their work is of high quality. Sending soil samples there (which means dealing with US quarantine) can seem daunting. But it really is not difficult. I do it myself when I can afford to wait a few weeks for a result. There's an appendix at the end of this book that provides the documents and information you need.

CANADA

I have not discovered any Canadian soil testing lab that routinely does Mehlich 3 extractions. I surveyed about 20 Canadian labs; several responded that they routinely do ammonium acetate extractions but can do M3 upon request. If it were my soil sample, I would not count on a lab's results when it was exploring new territory.

AUSTRALIA

Australians may use US labs. But I know of two labs in Australia that routinely use M3 extractions. At this moment, AgVita charges AU $103.40 per test. Contact:

- AgVita, PO Box 188, Devonport 7310, Tasmania, Australia. Phone: +61 3 64 209 600, agvita.com.au, email: dhicks@agvita.com.au. Ask for their "Lab 22 Complete Soil Test."
- Australian Perry Agricultural Labs, PO Box 327, Magill, South Australia 5072. Phone: 08 83320199, apal.com.au, email: info@apal.com.au.

So how accurate are soil test results? Well...quite accurate, but intentionally inaccurate as well. I say inaccurate because the M3 test measures the concentration or relative proportions of elements contained in a specific volume of topsoil. And when you fertilize in response to that information, you aim to create specific concentrations, which are the target levels. Ideally, it should be the *actual* weight of that topsoil acre that determines how much fertilizer that layer of soil should receive. If the top six inches of an acre actually weighs two million pounds, and if there are 2,000 pounds of calcium discovered to be in that layer of soil, then calcium occupies one-tenth of 1% of the total weight. But if that same volume of soil only weighs one million pounds, then the same quantity of calcium comes to two-tenths of 1% of the total — double the concentration. In most cases, the concentration is what's critical, not the amount. Inexpensive soil testing labs do not measure the density of the sample you send in and compute from that the true weight of the furrowslice acre. Instead, they assume by default an average value: 2,000,000 pounds per acre. Doing it this way greatly lowers the cost.

So does it really matter if the reported concentrations are off by 20% one way or another? Answer: yes and no. Yes, in that the amount you are told to apply will also be off by that amount. So you might be led to put in a bit more or a bit less than what a perfectly accurate test would call for. But the whole business of soil remineralization is pretty loosey goosey. The mineral profile of each sample that went into your mixing bucket was somewhat different. The average of these samples may not precisely match any specific point in the field. However, over time, the inaccuracy all works itself out in your favor. Suppose your six inches of topsoil actually weighs 2,200,000 pounds per acre; the soil audit assumes it is 2,000,000 pounds and instructs you to put in 100 pounds of potassium, but you actually needed 110 pounds to achieve the desired concentration. So that year, you may harvest a bit less than you might otherwise have. This minor yield shortfall will go entirely unnoticed by the home gardener — who is not counting crates being shipped off to market. Actually, gardeners will probably be amazingly pleased when comparing their yields to what they had been when the soil was lacking that 100 pounds. So, next year's test will still show some potassium deficiency. And maybe the subsequent test will call for only 40 pounds of potassium. Or maybe another 100 pounds — if the crop used all you fed the soil. My point, despite the inevitable inaccuracy, is that, over the years, you'll get to the same place — and you'll enjoy good winds and fair weather the whole way.

I also said M3 tests are accurate. These days, the extractant is assayed using *Inductively Coupled Argon Spectrography*. Once the extractant solution reaches the analyzer, the whole process takes only a few seconds per test sample; all the discovered levels are entirely accurate. So we get a near-perfect assay of a soil sample, but the result only approximates the field as a whole because it is based on a rough assumption of the field's bulk density.

There is some variability from soil lab to soil lab. Although the method for doing M3 extractions is precisely specified, no lab does it quite the same way. In practice, this means that if you send the same sample to two labs, you will get two somewhat different answers. Some labs are known for consistently reporting higher levels; some consistently report lesser figures. Again, this variation does not matter much *as long as you use the same laboratory year after year.* Exactly like the variations that result from intentionally miscalculating the weight of the soil slice being analyzed, these variations all work themselves out if you work with the same level of error year after year. When farm advisors switch labs, they have to recalibrate; this is usually experienced as being stressful. Once you start using a soil lab, I suggest that you do not go doctor-hopping.

Doing the Worksheet

Here is a real M3 soil audit of a typically leached soil commonly found in the cooler and well-rained-upon parts of the northern United States and southern Canada. We're going to analyze this report together and work out the soil amendments needed.

To make analysis easy, I've developed a number of worksheets. When I was first analyzing my own soil audits, I found it required lots of flipping back and forth looking for tables of numbers, figures and ratios. With my worksheets, the entire process will only take you a few minutes, and it's a breeze.

Before going further, please make a few photocopies of the blank *Acid Soil Worksheet* included in the Appendix. Or, a letter-sized version of all the worksheets can be downloaded from the New Society Publishers website, newsociety.com. Incidentally, in the event I change my opinion about what constitutes the best target levels (and I'm likely to), the downloadable worksheets will be promptly upgraded; any changes to the book must await reprinting — if ever that happens.

Soil Report

Job Name **Matthew** Date 2/3/2012

Company Matthew Submitted By

Sample Location			Unfarmed				
Sample ID			West				
Lab Number			50				
Sample Depth in inches			6				
Total Exchange Capacity (M. E.)			12.87				
pH of Soil Sample			5.80				
Organic Matter, Percent			5.36				
ANIONS	SULFUR:	p.p.m.	26				
	Mehlich III Phosphorous:	as (P_2O_5) lbs / acre	954				
EXCHANGEABLE CATIONS	CALCIUM: lbs / acre	Desired Value	3501				
		Value Found	3292				
		Deficit	-209				
	MAGNESIUM lbs / acre	Desired Value	370				
		Value Found	210				
		Deficit	-160				
	POTASSIUM: lbs / acre	Desired Value	401				
		Value Found	125				
		Deficit	-276				
	SODIUM	lbs / acre	73				
BASE SATURATION %	Calcium (60 to 70%)		63.93				
	Magnesium (10 to 20%)		6.80				
	Potassium (2 to 5%)		1.24				
	Sodium (.5 to 3%)		1.23				
	Other Bases (Variable)		5.80				
	Exchangable Hydrogen (10 to 15%)		21.00				
TRACE ELEMENTS	Boron (p.p.m.)		0.63				
	Iron (p.p.m.)		206				
	Manganese (p.p.m.)		8				
	Copper (p.p.m.)		5.16				
	Zinc (p.p.m.)		31				
	Aluminum (p.p.m.)		949				
OTHER							

Logan Labs, LLC

Fig. 5.1: *Logan Labs Soil Report.*

Here is what to do: Set the Logan Lab *Soil Report* and a blank *Acid Soil Worksheet* next to one another. On the *Soil Report,* look for the amount of each element the test discovered. The Logan form reports

Acid Soil Worksheet

U.S. Measurements

Name *Matthew*

Plot or Field *Unfarmed*

Date of Test *2/3/2012*

Sample Depth 6 inches All numbers on this worksheet assume a six inch sample depth

TCEC *13* **TEC:** If CEC is below 10 it is a "light soil." Over 10 is "heavy soil."

pH *5.8* **pH:** If pH is 7.0 to 7.6, go to the Excess Cations Worksheet. If pH exceeds 7.6, you may have calcareous soil.

Organic Matter % *5.4* **OM:** Target over 7% in cool climates. South of the Mason-Dixon Line target over 4%. Assume an approximate release of 15–25 lb nitrogen per 1% OM. Varies with temperature, moisture and soil air supply. N = 0.22 x NO_3

Element	Actual Level	Calculating Target Level Pounds per acre	Target Pounds per acre	Deficit
Sulfur **S**	ppm *26* lb/ac *52*	**S = ½ Mg (Target Level)** until there are no cation excesses; then **you *may* Target S=¹/₃ K**		
Phosphorus **P**	P_2O_5 *954* P = *420*	**P = K (Target Level)** Calculate using actual P, not phosphate. P = 0.44 x P_2O_5		
Calcium **Ca**	lb/ac *3292*	**TCEC x 400 x 0.68 = Target Level**	Minimum target 1,900 lb/ac	
Magnesium **Mg**	lb/ac *210*	**TCEC x 240 x 0.12 = Target Level**		
Potassium **K**	lb/ac *125*	**K is proportional to TCEC: see chart**		
Sodium **Na**	lb/ac *73*	**TCEC x 460 x 0.02 = Target Level** Be certain of good water quality before adding sodium	Do not exceed 160 pounds	
Boron **B**	ppm *0.63* lb/ac *1.3*	**B = 2 lb/ac if CEC below 10 = 4 lb/ac if CEC above 10**	Do not exceed 4 pounds	
Iron **Fe**	ppm *206* lb/ac *412*	**Fe = 100 lb/ac if CEC below 10 = 150 lb/ac if CEC above 10**		
Manganese **Mn**	ppm *8* lb/ac *16*	**Mn = 55 lb/ac if CEC below 10 = 100 lb/ac if CEC above 10**		
Copper **Cu**	ppm *5.16* lb/ac *10.3*	**Cu = ½ Zn (Target Level)**		
Zinc **Zn**	ppm *31* lb/ac *6.2*	**Zn = ¹/₁₀ P (Target Level)**		

Potassium Target Levels

TCEC	Pounds	TCEC	Pounds	TCEC	Pounds
		16	390	28	493
		17	400	29	500
		18	410	30	507
7	255	19	420	31	511
8	270	20	435	32	515
9	290	21	443	33	519
10	310	22	451	34	523
11	320	23	459	35	527
12	335	24	463	36	531
13	350	25	475	37	535
14	365	26	481	38	539
15	380	27	487	39	543

	One acre, six inches deep weighs	One hectare, 80 mm deep weighs
1 meq Calcium	**400 lb**	**400 kg**
1 meq Magnesium	**240 lb**	**240 kg**
1 meq Potassium	**780 lb**	**780 kg**
1 meq Sodium	**460 lb**	**460 kg**

1 ppm = 1mg/kg = 2 pounds/acre = 2.24 kg/hectare

If TCEC is lower than 7, use value for 7. If it is over 39, use value for 39.

Date of Issue: 07/07/2012

Fig. 5.2: *Filled in Acid Soil Worksheet.*

"ppm" for S and the trace elements (iron, manganese, copper, zinc and boron); use their "Value Found" in lb/acre for Ca, Mg and K and "lbs/acre" for phosphate and sodium. Enter these amounts into the column on the worksheet's left-hand side where it says "Actual Level." To make things easier, I've designed the *Acid Soil Worksheet* so it closely resembles the layout of Logan Labs' *Soil Report*. Other soil testing laboratories present their data in slightly different order; they may also give additional information that you do not need to deal with. Note that Logan Labs reports calcium, magnesium, potassium, sodium and phosphorus (as phosphate) in pounds in a furrowslice acre; the rest of the elements are given in parts per million. Spectrum Analytic reports all their levels as parts per million instead of as pounds per acre and when calculating TCEC Spectrum often adjusts it downward. I urge all new soil analysts to use Logan.

Where the figures on the soil audit are given as parts per million (ppm), first write in the ppm given, then multiply that amount by 2 to arrive at pounds per acre (because we're assuming a soil slice weighs two million pounds per acre). Then enter lb/ac in the lower half of the space provided. Where Logan reports on the major base cations — calcium, magnesium, potassium — notice that there are three numbers: "Desired Value," "Value Found," and "Deficit." Enter the "Value Found" on the worksheet. We will work out our own "Desired Value" and determine for ourselves any "Deficit" that may exist.

Logan Labs reports phosphorus as phosphate (P_2O_5), not as elemental phosphorus (P). Since everything else on this worksheet is calculated in pounds of the *elemental* substance, convert phosphate to elemental phosphorus by multiplying phosphate by 0.44.

The example soil audit is for a homesteader's unused bottomland acre beside a small creek in southwest Washington State, not too far from Portland, Oregon. It is alluvial sandy loam soil currently growing pasture grasses. I know from having lived in Cascadia myself that this highly leached soil gets about 80 inches of rainfall every winter. From the high level of phosphorus, it is my guess this land was used as a market garden sometime before Matthew purchased it.

Now, we're going to discuss the meaning of Logan Lab's soil report form, line by line.

Total Cation Exchange Capacity

The most important number on the form is the *Total Cation Exchange Capacity* (TCEC). Knowing the TCEC lets you compute the total weight of all available plant nutrients that the clay and humus in a measured amount of soil is capable of holding on to. The soil audit tells you how much of that capacity is already filled with plant nutrients and how much of it, if any, remains to be filled by fertilization. TCEC also suggests how stably and abundantly your soil is capable of feeding your crops.

Sometimes this measurement is slightly misnamed "CEC," or "Cation Exchange Capacity;" sometimes the same thing is termed the TEC (Total Exchange Capacity). There is a difference, though. TCEC is the sum of all the effective exchange functions, both the clay fraction and the soil organic matter. CEC is the exchange capacity of a particular pure substance, like a type of humus or a particular deposit of clay. Some types of clay could have a CEC of 100, but if it constitutes 33% of the furrowslice acre, then, assuming there is no organic matter at all, the TCEC of that soil will be 33. Don't be alarmed if this sounds too technical. TCEC is basically a valuable number calculated by the soil lab from the test results.

TCEC Feeds the Crop

Plants obtain nutrients from several sources. The most instantly accessible nutrients are those dissolved in the soil moisture, called the *soil solution*. Hydroponics is an extreme example of this situation. If a soil contains no clay and no humus, the soil solution would be the only immediate source of plant nutrition; in that event, we'd have to frequently side-dress soluble fertilizers in small amounts or fertigate with them already in solution. The other alternative would be to use slow-release fertilizers, which allow nutrients to dissolve continually, but gradually. We would have to fertilize constantly because no matter how moisture-retentive a soil may be, once a leaf canopy has formed, the crop will remove almost all readily accessible topsoil moisture in only a few sunny days, so if this hypothetical soil is entirely without an exchange capacity, plant growth stops once the soil solution has been pretty much used up. On the other hand, if it rains hard enough to move water from the surface down through more than a foot of soil, this hypothetical soil will be leached of nutrients, and plant growth will screech to a halt until the soil solution is recharged.

Of course no growth is the last thing a grower wants. A non-growing plant is a stressed plant. And a stressed plant becomes a disease- or insect-attacked plant. And that plant will fail to produce the best tasting, most nutrient-dense harvest. You want your plants to grow steadily — without interruption and without stress and abundantly supplied with every useful nutritional element.

If the soil solution were a plant's only nutrient supply, you'd be practicing something like outdoor hydroponics. And, in truth, many farmers are doing nearly that. They monitor the fertility of the soil solution almost from day to day and send tissue samples off to the lab to discover what elements the crop is actually managing to uptake (as opposed to what the soil test says should be available). Then they inject ever-changing fertilizers into each irrigation. My own experiments involving large containerized plants in an artificial growing medium (meaning the plants were feeding exclusively from the soil solution) showed that this method is *not* an easy, uncomplicated, trouble-free way to grow things. And judging by what's offered on the local supermarket fruit and veg counter, soil-solution growing does not produce nutrient-dense foods.

Plants grow best when a far greater abundance of nutrients than could ever be dissolved in the soil solution are lightly attached to the surfaces of clay and/or to humus. Attached nutrients become available easily, but they are not dissolved in the soil water. It's much like how power is available to flow from a battery. Here's how it works: Clay is capable of holding onto cations electrically (similar to static cling). It also releases cations to plants on demand. This release is termed an *exchange*; the ability to exchange lightly attached nutrients is termed the *cation exchange capacity*. Humus is more effective at making this sort of exchange than clay is. It's better in two respects: it holds and releases both cations and anions, whereas clay holds only cations; and, gram for gram, the better sorts of humus can hold and release two to four times more cations than the very best type of clay can. Cations and anions that have stuck themselves on clay or humus are not in the soil solution but they stand ready to immediately replace whatever plants remove from the soil solution. Attached cations and anions do not readily leach out with hard rain or overwatering. The most effective way I have ever heard the nature of TCEC expressed in simple language was this: if the plants are feeding at the dining room table, the nutrients in the soil solution are like the food on their plates. The TCEC is food in the pantry, ready to be brought out and put on the table as needed. The bigger the pantry, the longer the dinner can go on.

Without importing soil fertility, the source of nutrients is the slow, ongoing breakdown of soil mineral particles, what Albert Howard termed the *annual increment of fertility*. Every soil has a unique annual increment. For the great majority of soils, the increment's size is entirely insufficient to send abundant crops out the farm gate year after year — not even close. This is why fertilizer is used. But the trick is to figure out how much more fertility is needed than is produced by the soil's annual increment. By using a laboratory procedure considerably more costly than the standard Mehlich 3 test, the soil's entire reserve mineral content can be accurately measured, and from those numbers it is possible to make an educated guess about that soil's annual increment. In Chapter 2 I included some statistics showing typical total reserve nutrient capacities of some United States soils in different climates.

A total mineral analysis answers the question: Does a soil actually possess the basic nutrients to provide amply for crops, even though these reserves are not available now? If it does have the nutrients in reserve, it might pay to go to work on biologically liberating them at a higher rate. Or, perhaps a soil is so entirely devoid of some nutrient (or many) that there is no hope of bringing up the level, short of importing it. This second scenario is the most frequent case.

In good cropping ground, about half the soil's volume consists of solids — mineral particles and the clay that formed from minerals already dissolved. Air and water should make up the other half of the soil volume. There'll also be a few percent of organic matter and soil microlife.

Clay forms in soil from the remnants of dissolved soil minerals. Its nature varies according to the sort of rocks that originally decomposed to form it, as well as the climate it formed in and how long it has been

The first farm advisor I ever met was named Will Kinney; he operated in southern California. When I was a novice gardener Will taught me more in one afternoon's conversation than I've learned from reading many a book. Will bragged about his greatest farming success. It happened on a leased field in California's central valley. He negotiated a very low rent because the land looked impossible to farm. It was light yellow, wind-blown sand entirely devoid of organic matter, and it had thin salt deposits on the surface. But Will had that salt analyzed; it wasn't sodium, but a broad range of plant nutrients left over from previous fertilization. So he plowed the salts under and began regular, light irrigation to keep the topsoil moist. Within weeks, the yellow sand turned black from self-created organic matter that appeared in response to the presence of abundant mineral nutrients and water. For two years, Will took bumper vegetable harvests off that field without having to put in any fertilizer.

in place on a geological time scale. There's an excellent (and not too hard to follow) explanation of this in Foth and Ellis's *Soil Fertility*. Of most concern to gardeners is the capacity that the clay has to hold on to cations. Clay's cation exchange capacity varies greatly by type of clay, so overall TCEC varies according to what percentage of what sort of clay is in the soil. Old clays have lost most of their ability to hold cations. The TCEC of a geologically old, worn-out, nearly pure clay soil might be 4. A clay loam, usually the most productive sort of farm soil, might have a TCEC of 40 or higher, or it might be 25, or only 5, all depending on the nature of the clay.

I mentioned that humus has a big exchange capacity; in fact, the lower a soil's TCEC tests initially, the more important increasing soil humus becomes. Measure for measure, pure humus has a far higher CEC than pure clay. The weakest forms of humus have a CEC of about 100 units (I'll define "unit" later). If the soil were to consist of one-third of that sort of humus by weight, and had zero clay, you could expect to find a TCEC of 33 from the humus alone. Of course, no ordinary soil is one-third humus. Even compost gardeners rarely build more than 10% organic matter, and much of that 10% will not yet have decomposed into humus. The strongest humus has a CEC of about 400 units. The best possible clay has a CEC of about 120 units; the weakest clay, less than 5. It is not uncommon to find a sandy farm soil with so little remaining humus that it has a TCEC below 5. And it is not uncommon to find a clay-loam garden soil with an organic matter content around 10%, with a TCEC of 35.

Light Soils; Heavy Soils

The terms "light soil" and "heavy soil" are sometimes used to refer to the physical density of a given volume of soil. In terms of density, gold has a far greater density than aluminum. Looked at that way, a clay soil is a denser soil, and sandy soils are less dense. But when considering remineralization, "light/heavy" refers to the weight of cations the soil can hold on to. By definition, a light soil has a TCEC below 10; a heavy soil has a TCEC over 10. To get a practical feel for what TCEC means, look at it this way: to grow an average farm crop from start to finish requires feeding it all the plant nutrients that could be held by a TCEC of 7. Growing a crop on a soil having a TCEC of 7 just about empties the pantry by the time the crop has finished. Given a winter to rebuild (from the further breakdown of its organic matter or from the annual increment), a light soil's pantry will be partially recharged by spring.

A heavy soil has a voluminous pantry that can hold enough plant nutrients for several crops. We measure the existing content and size of that pantry with a soil test and then — when needed — we replenish that pantry to bursting with the right stuff in the desired relationships to one another. This allows our plants luxurious consumption of all important nutrients — in other words, they'll grow like crazy, be nearly immune to most forms of insect attack and disease, and provide us with nutrient-dense food.

Try as we might, we can't put enough nutrients into the pantry of a very light soil. To improve this situation, you can raise the soil organic matter level by incorporating humus. If you put in quite a bit, you can significantly up the exchange capacity. And this is really the point of adding humus. Many gardeners do not realize that the soil ecology does almost as well at 3% organic matter as it does at 5%; in light soil the main benefit of elevating organic matter is to raise the exchange capacity. Gardeners with light soil are better off using slow-releasing organic fertilizers to deliver a relatively steady nutrient supply over a long period. Farmers can rarely afford organic fertilizers or to buy, haul and spread manure (which is Nature's slow-releasing fertilizer).

With a heavy soil, the strategy is a bit different. You only need a moderate amount of compost, not the larger quantities it takes to lift TCEC in a light soil. Unless you're growing on a dense, airless, pure clay, you need add only enough organic matter to abundantly feed the soil's ecology. Once you spread fertilizer, stocking its pantry to bursting, you can then expect to successfully grow an abundant crop (or even several crops in succession during a long growing season) and not be much concerned about plant nutrition until it's time to retest and restock the pantry for the next spring. After a few years of this, you may come to see the unique pattern of your heavy soil's strengths and hungers, and so be able to anticipate its fertilizing needs for a few years without doing an annual test.

Clearly, it is easier to grow nutrient-dense food in heavy soil than in light soil. To get nutrient-density from a light soil, you need to greatly lift its humus content. You can do that most effectively with compost made in such a way that it develops the highest possible cation exchange capacity (see Chapter 9 on making compost). Great compost can elevate a lightweight soil into a middleweight contender.

Assuming there actually is one foot of topsoil to work with, an instant method to improve the performance of a light soil is to sample it 12 inches deep, calculate (double) quantities of fertilizer appropriate for that soil depth, and dig them in one foot deep, something most spades are capable of, if you are. That's a serious suggestion! It doesn't really make a heavy soil out of a light one, but it does make it *act* more like one. If you take this road, do not dig in *compost* a foot deep; dig in the fertilizers first and spread the compost on the surface. Mix the compost in with a rake, not a spade. Next time you dig, you'll turn

... I have discovered that my garden TCEC is 8 after many years of manuring and composting, compared to neighbor's 200 yards away CEC of 4.4 using commercial 10-10-10 and tilling in all residues. Which to my way of thinking is significant.

— John Weil, private correspondence.

under what's left of that compost; best to let it initially decompose near the surface, where there's the largest air supply.

Many organically approved fertilizers are insoluble (lime, greensand, phosphate rock) or are slowly soluble (gypsum and K-Mag); these are best dug in and thoroughly dispersed throughout the soil prior to sowing, although gypsum and K-Mag will work (less rapidly) when no-tilled. Expect coarsely ground garden lime, greensand and hard rock phosphate to take many years to break down. Some fertilizers, like the sulfates, are as soluble as table salt and work excellently when spread on the surface, as long as they are watered in.

On light soils, it is wise to anticipate the pantry running out before it actually does. About halfway through the crop cycle, plan to side-dress an additional half-dose of any soluble fertilizers called for by the soil analysis, plus a half-measure of oilseedmeal (or fishmeal, if you don't mind the temporary odor), much as you would side-dress with COF. These materials release effectively when surface applied. The sulfate salts leach in when it rains or the garden is sprinkler irrigated. The seedmeal's release happens because (micro)soil animals come to the surface to feed on it when the soil is moist.

You should side-dress before a crop forms a leaf canopy, or when crops like cucumbers or unstaked tomatoes have spread over about one-third of the ground they will ultimately cover. If you wait longer, it requires painstaking care to spread fertilizer without causing damage.

A crop may initially grow great in a light soil, but run into major imbalances and/or shortages because plants can be selective about which elements they assimilate. If the plant's withdrawals throw the soil strongly out balance, growth screeches to a halt. The plants get stressed and may become diseased or get attacked by insects. Boosting the soil's organic matter level does a lot more than add nutrition that gets released as the organic matter decomposes. And it does more than merely feed the soil ecology. Compost stabilizes the ups and downs of the soil solution. In light soils, this factor is much of what makes compost improve plant growth.

So. The biggest question to ask about a soil is this: Does it hold enough nutrients on the clay and/or on the humus to allow you to load up your exchange capacity with the right balance of nutrients, and then, with those nutrients safely in storage, grow a healthy abundant crop from start to finish without adding more (or very much more) fertility; or is the exchange capacity low, such that you are forced to provide an ongoing resupply of nutrients, either from small quantities of soluble soil amendments (what many farmers do) or from slow-releasing organic substances that have the ability to supply plants for several months from a single application (as wise gardeners do).

A *light soil* has a TCEC below 10.

A *heavy soil* has a TCEC above 10.

Light or heavy, the top 6 inches of most soils weigh around 2,000,000 pounds per acre, or 2,000,000 kilograms per hectare. If the soil's density causes the furrowslice acre to be much heavier or lighter than 2,000,000 pounds, gardeners can afford to ignore the difference.

The Arithmetic of TCEC

A powerful microscope reveals that clay is made of thin plates or sheets, stacked in layers. Around the outside edges of the clay layers are enormous numbers of negatively charged attachment points, better termed "exchange points." Like magnets with negative poles sticking out, each of these charged exchange points is capable of holding on to a positively charged atom — a cation. The clay/humus will uncritically accept any cation — be it calcium, zinc or uranium. The number given as the TCEC stands for *the quantity of negatively charged exchange points existing in a given weight of soil.* The total weight of all cations attached to those points is the total load of potential plant nutrients on the clay-humus content in a fixed depth amount of soil.

In the chemistry lesson a few pages back, please notice that next to the symbol for the first group of plant nutrient cations is a small plus sign; some have more than one plus sign. Some cations attach to clay or humus at only one charged point; some attach at two. It works something like that child's game of paper, scissors, rock: a cation with two attachment points hovers around a bit of clay, knocks off a pair of cations with only one attachment and replaces them or knocks off another cation with two weaker attachment points and replaces it. A cation with two strong attachment points, like calcium, is rarely replaced by a pair of cations with only one positive charge, like potassium. Usually, most of the TCEC will be saturated with calcium and, to a lesser degree, with magnesium, because both of these elements have two attachment points, and these two elements are naturally plentiful. Calcium attaches more strongly than magnesium can.

Cations are tiny. Whenever a crystal of table salt dissolves, jillions of them get liberated. Suppose the TCEC number really means there are so many hundred jillion attachment points in a furrowslice acre. It seems a huge number, but we're talking about atom fragments. Individually, they don't weigh all that much, so for convenience we measure them collectively in pounds per acre or kilograms per hectare. We load up the soil's exchange capacity with a fixed number of cations that we measure out as pounds or kilograms of soil amendments. Add any more cations than the number of attachment points to hold them, and the surplus cations just hang out in the soil solution — where they can be easily leached out.

Now, here's where the conversation gets a bit sticky for people without a good grasp of arithmetic. Total Cation Exchange Capacity is expressed as *milliequivalents*, or meq. A milliequivalent is a specific

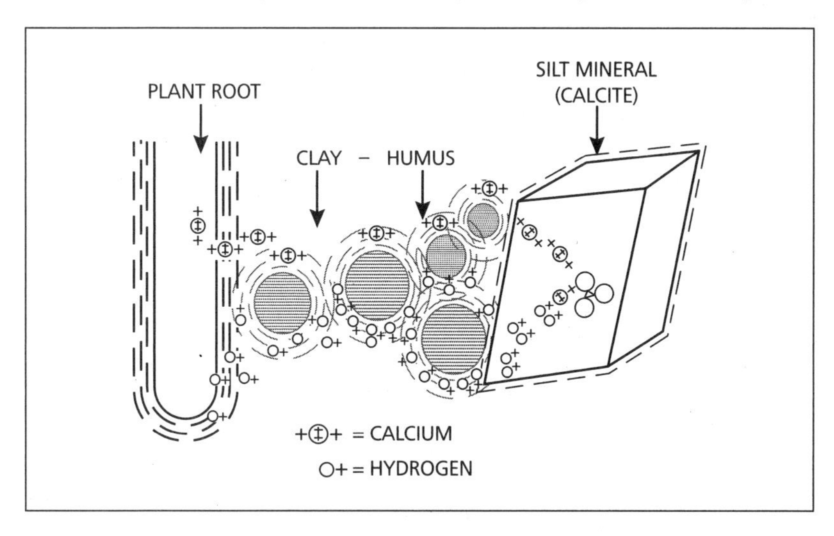

Fig. 5.3: *Cation exchange capacity illustrated.*

number of attachment points present in a certain amount of soil. When talking about the general theory of it, CEC is defined as milliequivalents per 100 grams; on that scale, a milliequivalent of calcium would be a fraction of one gram. When growing crops, we want to know the weight of one meq of any particular element in an acre of soil of a particular sampling depth. Usually this means the furrowslice acre. To work with a soil test report, you have to manipulate that number, so it's best to understand what it's all about. Don't worry. I'll explain.

Please contemplate the numbers in the table. They express meq in pounds per acre because humans cannot quite imagine one billion cations, far less one jillion; we simply can't envision what those long chains of zeros really mean. But we can grasp pounds — 50 pounds or 5,000 pounds — and we can get a reasonable grasp on the mass of a thing. So, when working out a soil analysis, we do not consider that a teacup-full of soil contains 6,240,400,000,000 cations of calcium and 1,765,844,000,000 cations of potassium. Instead, we deal with the *weight* of calcium, magnesium, potassium or sodium. So, in working with a soil test report, we are concerned with the weights of the elements already present as well as the weight of what we want there ideally. You must use meq when working out the weights of calcium, magnesium, potassium and sodium from your own soil test result, but there is no need to remember the millequivalent weight of each element or even to remember that these weights are printed on this page; the numbers are also on the worksheet, which is where you'll need to use them. Here's how the math goes:

Atoms come in a range of sizes and weights. Calcium atoms weigh more than magnesium atoms do. And potassium has a greater weight

	One acre	**One hectare**
	6 inches deep	**150 mm deep**
1 meq calcium	400 lb	400kg
1 meq magnesium	240 lb	240 kg
1 meq potassium	780 lb	780 kg
1 meq sodium	460 lb	460 kg
1 ppm =2 pounds/acre=2 kilograms/hectare		

Table 5.2: *Weight of One Milliequivalent.*

than sodium. If one meq (millequivalent) of the total exchange capacity of a furrowslice acre were completely occupied by calcium, those jillions of cations would weigh 400 pounds. If that furrowslice had all its exchange points filled with 1 meq of magnesium, that acre of topsoil would hold 240 pounds of magnesium. So, if we have a soil test result that gives a TCEC of 7.0, it means that the furrowslice acre is capable of being completely saturated by 7 milliequivalents of calcium (7 x 400) which equals 2,800 pounds of calcium; or it could equally be saturated by 6 meq of calcium (6 x 400) and 1 of magnesium (1 x 240) which equals 2,640 lbs of cations; or it could hold 5 meq of calcium, 1 of magnesium and 1 of sodium, totaling 7 meq (and weighing?...you calculate that; it's a little test for you). In each case, the total weight of cations would be different, but the number of attachment points involved would be identical — 7 meq.

The System

Now that you understand TCEC, I can better explain the basic strategy for producing nutrient-dense food. The goal is to provide the highest levels of plant nutrients that the soil's exchange capacity can hold on to, targeted so as to be in a pre-determined balance, one with the next. As TCEC rises or falls, the amounts of all (or most) plant nutrients we add rise and fall accordingly. Sounds simple, but actually doing it rarely comes cheap.

Full soil remineralization is not affordable for most farmers under current economic circumstances. Farmers feed just enough of the

To grow an average farm crop requires roughly the amount of plant nutrients that can be held on clay and humus with a total cation exchange capacity of 7 meq. Therefore, it is workable strategy to fertilize any soil with an exchange capacity below 7 as though it could hold 7. However, if we apply that much fertilizer all at once, some of it will not stick to the clay, and we'll have to take care not to overwater. And hope it doesn't rain too much. And hope that if some of the fertility does leach downward, that the next 6-inch layer of soil will be capable of holding on to it. You can see why, when gardening in light soils, it's best to use materials that release slowly.

cheapest fertilizers that will get them peak yield, not high nutrient-density. Few farmers would ever put in much more than that, especially not indebted farmers. In conventional agronomy, the amount of fertility that brings about peak yield is termed the "strategic level." The level of concentration at which any further increase would not make any more yield is called "sufficient." In commercial farming, the whole game is to achieve sufficiency. Biological or holistic farmers and gardeners use a different approach. The major cations — calcium, magnesium and potassium — are balanced according to exchange capacity. The other nutrients are variously handled, sometimes balanced, sometimes brought to levels beyond strategic. Farmers who use cation balancing often produce the sort of nutrient-density this book encourages.

Perhaps you wonder if my book's targeted levels are spot on the bull's-eye. In truth, I doubt they are. I know there cannot possibly be only one bull's-eye for all crops and soils: soils are too variable; crops differ widely in their needs. But targeting a balanced abundance works a lot better in a garden than anything else I know of. And it works for any gardener who can do arithmetic or enter values into a spreadsheet. For sure, balance increases nutrient-density, and it does it a lot better than the SaMOA system. There are a dozen prominent soil analyst-authors out there, each of whom holds somewhat different opinions about what the ideal soil would be. If you delve deeper than this book, you'll have to make some choices yourself.

I caution you about delving. First, different agronomists use different sorts of soil tests. If you plug numbers into my worksheets that were derived from another form of extraction...well, it won't work so well. As far as I know, anyone in any state or province can legally advise farmers about their soil without possessing formal qualifications or a license. I know of several radical farm consultants whose background is in holistic medicine (human or vet), and they have no formal education in agriculture. Farm consultants who write books are self-educated more often than not; some are highly idiosyncratic and/or creative about using novel terms and unannounced redefinitions. Sometimes, you have to interpolate spaces between words to get their drift. Farm consulting also involves serious money. Twelve hundred dollars a day plus all expenses is not an unreasonable rate to pay a top-flight consultant. Against a multi-million dollar farming operation, a few days of that consultant's time is small change. With such a comfortable standard of living at stake, consultants seek to distinguish *their* method as the best method — THE best method. To me, many self-help agronomy books are mainly advertisements for the consulting services of the agronomist who wrote it.

Erica Reinheimer and I put in long hours over six months of research assessing that entire body of farming-advice literature, comparing the preferred nutrient levels of numerous soil consultants as we worked out safe, effective targets that achieve nutrient-density in the home garden. To that body of information, we added 40 years of my gardening experience and another 30 of Erica's, including several years of her practice as a soil analyst. You may be the sort of person who is compelled to reinvent the wheel; if that describes you, then you are invited to reconsider our research and draw your own conclusions; the Bibliography will lead you to this information. However, I suggest that for now, you accept our targets and procedures. Using this book, grow your greatest garden ever for the next few years and then see if you want to learn more.

Matthew's soil has a TCEC of 13. It is a heavy soil.

Pee Haitch

The cation-balancing method separates soils into two basic types according to pH: acidic or neutral/alkaline. (There's a sub-type of alkaline soil, calcareous, a soil that is alkaline because it contains very large quantities of calcium.) The pH level on the soil test report will direct you to the worksheet that is appropriate for your basic type of soil. Because the greatest portion of North American agricultural soils are naturally acidic, this book will first focus on how to fill in the *Acid Soil Worksheet.*

Forty years ago, mainstream agronomists made a huge thing over soil pH; some still do. Not too many years ago, the pH test was supposed to be the only soil test a farmer or home gardener needed. Some soil analysts still hold that opinion. Testing soil pH is a cheap, simple procedure; you can get a rough answer from a bit of fish-tank litmus paper. Farm advisors-in-training learned that acidity was undesirable; pH should be raised by adding finely ground limestone. I've seen easy-to-use tables for gauging lime that say things like: To move a sandy soil from pH 5.5 to pH 6.25, add so much lime; add quite a bit more to shift a clay soil the same amount of pH. Etc. I suppose liming by pH is better than not liming at all, but the method has a catastrophic flaw.

In the realm of home gardening, where almost any damn-fool idea can be revered for decades, it was often suggested that since lime is necessary, the gardener may as well use dolomite lime, which is calcium-magnesium carbonate, $CaMg(CO_3)^2$. You supposedly get two advantages by using dolomite for pH adjustment: one ton of high-magnesium lime raises pH more than one ton of high-calcium lime. (Some of that increased effectiveness happens because 1 meq of calcium weighs 400 pounds; the same quantity of magnesium weighs only 240 pounds, so calcium-magnesium carbonate gets you more cations per ton than straight calcium carbonate.)

Magnesium is also an essential plant nutrient, so by spreading dolomite you get two nutrient elements for the price of one. However, this advice caused a lot of grief for a great many gardeners who were working soils already rich (or too rich) in magnesium. By using dolomite, they inadvertently created magnesium excesses.

Recall those jillions of negatively charged points on the clay. Every one of those charged points *must* have some sort of cation stuck on it. Absolutely must! If no other cations are available, hydrogen will slip in because an unfilled negative attachment point will rip a hydrogen cation out of the soil water to satisfy its hunger. But all other cations stick to clay harder than hydrogen — and they will replace hydrogen. If the soil solution has enough other cations in it, there will be no hydrogen cations. And, in that circumstance, the pH will test 7.0 or higher because pH is defined as the "density of hydrogen cations in water." The term pH itself means *potential hydrogen*. On the pH scale, 7.0 is the midpoint, where there is zero exchangeable hydrogen present. The scale increases in both directions, from the center. Below 7.0 is acidic (hydrogen increases in concentration as the number declines); 7.0 is neutral, there are no hydrogen cations. Above 7.0 is a scale of increasing alkalinity. To give you a practical feel for it, 5% household vinegar has a pH of about 3.2 and lemon juice usually about 3.0, lead-acid batteries are at 1.0. In the alkaline direction, baking soda is 8.3, and household lye is 13.0.

Actually, soil pH below 6.4 is not necessarily a result of too little lime or of too little calcium. It is a result of the soil's exchange capacity holding too many exchangeable hydrogen cations and too few calcium, magnesium, potassium and sodium cations. Adding any of these cations increases soil pH (i.e., reduces the number of hydrogen cations being held). Having a soil pH above 6.4 does not mean there is plenty of calcium and no lime is needed. It means that there are enough cations of magnesium, potassium, sodium and calcium *combined* to have eliminated most or all the hydrogen cations. It is often the case that a soil will test mildly alkaline, have a pH of 7.5, yet be shockingly short in calcium. In that soil, nothing grows well.

Lime is not the entire answer: adding high-calcium lime according to a pH test does not remedy having too few potassium, sodium or magnesium cations on the clay. To produce nutrient-density, you have to bring the four major cations into a particular balance: on naturally acidic soils, that balance will be 68% calcium, 12% magnesium, roughly 4% potassium, and 2% sodium. This is often simply put as 68:12:4:2, When soil is at that balance, its pH will be 6.4, which is our target pH for most soils.

High pH can also indicate there is an excess of any of the other three major cations; these excesses are not so innocent. Gardens irrigated with water carrying more than a small bit of sodium can show high pH for this reason. I'll soon have much to say about sodium and the great harm it can cause when in excess. There are also high-pH soils because they come with a built-in excess of magnesium. And there can be potassium excesses; these are usually found on farm soils that got too much chemical fertilizer and not nearly enough ag lime. High-potassium soil can be highly deficient in calcium and won't grow crops well until it gets plenty of lime — yet its pH can be above 7.0.

This is why I suggest that liming to adjust pH doesn't always work. The best book I've ever read about this is Victor Tiedjens's *More Food From Soil Science.*

Matthew has naturally acidic soil. It's pH is 5.8

Soil Organic Matter

Humus is a stable substance highly resistant to biological decomposition. Humus *may* be a long-term resting point for organic matter decomposition; sometimes the end point is complete consumption after a few months; nothing is left. The percentage of soil organic matter reported on a soil audit can be either stable humus or rapidly decaying organic matter. Both are essential. Chapter 9 explores the crucial difference between the two.

The presence of soil organic matter creates many positive effects. It stops erosion, increases water infiltration, builds pore space (soil air), etc. In fact, without humus, soil changes from being a living entity into dead dust ready to blow in the wind. But for now, please focus on how soil organic matter influences the soil's exchange capacity. Humus behaves as though it has an *anion* exchange capacity; without soil organic matter, plants cannot receive a stable abundance of the anion nutrients — sulfate, phosphate, nitrate and borate. Additionally, the soil microorganisms that are actively eating organic matter must build their bodies (mostly the proteins) using anions. Without soil organic matter and the complex soil ecology it feeds, anions (except for phosphorus) would wash out with the first decent rain.

Soils lacking humus cannot assimilate large additions of anions. Add too many, and they leach (or lock-up and become unavailable in the case of phosphate). Allow that to happen, and at the very least you have wasted money. This is one reason to limit the quantity of boron, sulfur and phosphorus added at any one time. If the generous application limits I recommend in this book do not result in even slightly higher numbers in next year's test, you probably need a higher organic matter content. Rather than broadcasting anion fertilizers, you could incorporate them into the compost heap to be sure they have married with the microlife. If anions seem to disappear into your soil, you can take comfort in this: even temporary high levels of phosphorus and sulfate build an expanding spiral of ever-more soil organic matter to hold more anions.

The blade cuts both ways: additions of compost adds more soil organic matter that holds more anions. If the soil test indicates organic matter is low, start importing it in abundance. If it then tests in a good range for your climate and soil type, you can relax. Hopefully, the amount of compost the garden itself generates will maintain that level once the soil gets into proper balance — or at least come close to being enough.

The ideal level of soil organic matter varies by climate: the warmer the climate, the more difficult it is to build a high percentage of soil organic matter because decomposition happens faster. If you're gardening south of the Mason-Dixon Line, consider yourself lucky if you can raise soil organic matter levels to 4%. You should strive for 5%, but you may never succeed. I have frequently seen garden soils with 10% organic matter in the northern tier of states and those parts of southern Canada where most people live. But 10% may be excessive. Honestly, folks, 4% in the South and 7 or so percent in the North is a gracious plenty.

I suggest you do not dig in enormous quantities of compost or semi-decomposed manure all at one go, except when starting a new garden. If you do spread it thick on new ground, give the soil a few months (or an entire winter) to digest such a heavy application before sowing seeds. Generally, a skimpy scattering of *high-quality* compost — a layer only one-quarter-inch thick — *maintains* a high organic matter level and perhaps slowly increases it. No matter the climate or soil type, to increase soil organic matter fairly rapidly, spread twice that amount — a layer of *high-quality* compost one-half inch thick. This is the largest amount of finished compost you should routinely spread. If you wonder at my italicizing "high-quality" twice in this paragraph, you'll understand when you read Chapter 9.

Improving the soil's mechanical properties, increasing its air supply, eliminating crust formation, facilitating seed germination, etc., have long been accomplished by building high organic matter levels. This approach was the best method we had before we found out about balancing calcium and magnesium. Building super-high organic matter levels as the solution to all ills is still being energetically promoted. True, build enough humus and you can counteract the way excess magnesium tightens up soil (and I'll soon have much to say about magnesium). But balancing the calcium-to-magnesium ratio will loosen that soil quicker and better, and the results can be much longer-lasting. Balancing is better than staying on that exhausting treadmill of hauling and turning and heaping and spreading. Hauling and spreading compost works, but doing that will certainly make *you* work too. But more importantly, importing large amounts of organic matter can bring with it large quantities of minerals, needed or not, to throw your soil off balance.

So, if your soil test indicates you have enough organic matter, my suggestion is to ease off on the imports. However, you should always recycle the garden's own plant waste, if for no other reason than this feedstock makes the best possible compost for delicate crops like cauliflower and celery — or any crop that has a history of not growing well on your ground. Should this year's test show your organic matter percentage has dropped from last year's, and is now below 4% on light soil or in the South, or under 7% on heavy ground or in the North, it's time to start building it up.

For a great many years, my books have *incorrectly* asserted that without imports, there is no way a food garden can develop a higher percentage of organic matter or even maintain its own current level. All my experience said that compost made from the garden's own waste should satisfy about one-third of the garden's annual need for humus. I expected to import enough biomass to manufacture about two times the amount of compost that the garden produced itself, or else import compost itself. This was the actuality of my own practice over the last 15 or so years. It's been the reality of SaMOA market gardening going back at least to 1850.

I've learned a few thing lately and have come to believe a fully remineralized garden soil can become a closed system in terms of organic matter. The entire biology of a balanced soil develops more...I have to call it *energy*. A highly mineralized balanced soil starts manufacturing a great deal more organic matter all by itself.

The forms of clay that develop in heavy soils usually have higher exchange capacity than the kinds of clay found in light soils. So, in accord with that difference, I expect humus forming in heavy soil to have higher exchange capacity than humus forming in light soil. In short: a balanced heavy-soil garden has the potential to become a self-sufficient entity quicker and easier than a light-soil garden. However, if gardeners with light soil import a bit of high-CEC clay into their compost heaps, they can produce humus as powerful as that found on heavy ground.

Humus

What is humus? Truth is: chemically, we still don't know. Humus is highly variable and nearly impossible to analyze in a lab. A good operating definition for humus is *the resistant bits remaining after all the easy-to-rot stuff has rotted.* I consider humus a form of activated or potentized clay because clay is an intrinsic part of humus. Humus will not form without some clay being present; should there be no clay, then organic matter rots away to nothing in a few years. But when the later-stage decomposition materials combine with clay, they form something almost as resistant to further rotting as clay is itself, and this material is called humus. Humus can remain in soil for hundreds of years.

Potters know clays vary greatly in their mechanical properties. Pure clay can have a CEC ranging from below 5 to above 100; when young (geologically speaking), clays are powerful magnets that attract many cations; as they age, their CEC declines. Pure humus can have a CEC ranging from 100 to 400. Why so variable? I speculate this difference develops from the nature of the clay involved when the humus formed. If you are gardening an old, weak clay soil with a low CEC, your composting produces lower-CEC humus. If you garden on sandy ground containing next to no clay and put this soil into your heap, you'll end up with next to no humus and experience a very large reduction in volume from what you started with. If you put soil containing a high-CEC clay into your heaps, they'll produce a high yield of high-quality humus.

In a nutshell, that's why light soils, including weak clay soils, do not develop high organic matter levels.

In Matthew's cool climate the soil organic matter should be 7% or more.

Organic Soils

Some rare (and horticulturally valuable, if they're drained) soils form in swamps of peat and other preserved organic matter. Amending "muck soils" effectively is not as simple as handling a mineral-based soil. Any time a soil test shows more than 15% organic matter, you probably have peat or something similar, and should consult local expertise about its care and handling.

The Four Major Cations

Let's return to the Logan Labs' *Soil Report*. Their standard soil test first lists two anions, sulfur and phosphate. But I'll begin with the four major cations — calcium, magnesium, potassium and sodium. Bringing these four elements into balance is the most important task in remineralizing. Heavy acidic soil may require more milliequivalents of calcium than a soil should usually be asked to accept all in one application. So it may take you a few years to bring a very heavy soil into balance. However, you should not be discouraged by the number of years involved because each step you take will make a big improvement as the proper balance is approached.

When calculating levels on the worksheet, always begin with calcium and magnesium.

Calcium and Magnesium

Calcium is an incredibly important plant nutrient much undervalued by gardeners. Crops do not intake nearly as many pounds of calcium per acre as they do some other elements, but crops do not grow without it being present in much greater quantity than all the other elements combined. Calcium is naturally abundant and is the most forceful element at attaching itself to the exchange sites; the TCEC is normally more saturated with calcium than any other element. But it is not so much the plant-nutrient aspects of calcium that concern us right now; it is how it relates to magnesium and thereby determines the soil's mechanical properties.

Calcium-to-Magnesium Saturation

I can hardly find words to express how pleased I feel when walking on the spongy carpet my garden soil became after bringing calcium and magnesium into balance. Two years ago, I had excess magnesium; I could hardly work up a fine seedbed in the spring without making clods; trying to dig after winter rains was exhausting. Then I boosted calcium without adding more magnesium. Now, my clay loam crumbles beautifully, even when it is wet. No more clods. Two years ago, I routinely unloaded mushroom compost, 200 bags at a time, and dug ten of them into every 100-square-foot bed once a year in an effort to keep the soil loose and open. Now, I don't think I'll need to import 'shroom compost again, or certainly not nearly as much of it.

The amount of calcium compared to the amount of magnesium on the exchange points determines if the soil is open, airy and loose or if it is tight and airless. It determines if the clay portion of the soil clings tightly to itself or if it opens up and separates — *flocculates* is the technical term for this. The ratio of Ca to Mg has as much or more effect on the soil's air supply as the level of organic matter does. Ever since J.I. Rodale captured most of the organic movement, garden books and magazines strongly urged loosening and aerating soils by building humus. The early garden writers did not know that when Ca:Mg is in balance, soils require *a lot less* compost. A well-flocculated clay or clayey soil is naturally open, loose and sponge-like and does not compact easily. An otherwise identical soil that has more magnesium on the TCEC, gets tight and airless, develops poor drainage and sticks to your boots. You shouldn't drive or walk on it when it's wet, or it'll turn into rocks.

A productive soil must consist of about half solids — mineral particles like sand, silt, clay and humus. The other half of the soil volume should be a constantly shifting balance of air and water. The soil gets wet; water displaces soil air. The soil dries out; fresh air must be drawn into the soil to fill the gaps. Roots breathe in oxygen; they exhale carbon dioxide. So does the soil biota — breath oxygen and exhale carbon dioxide. The reduced pore space and low rate of air exchange in a compact soil makes soil air too rich in carbon dioxide and too low in oxygen. In this inhospitable environment, plants fail to grow or grow less well. Without a plentiful oxygen supply, all the wondrous things the soil ecology can do happen at a much lower level — if they happen at all. Crops growing without enough soil air often become diseased or insect attacked. Consequently, Logan Labs' M3 standard soil test targets a 68:12 base cation saturation ratio: 68% calcium and 12% magnesium. The preponderant, overwhelming majority of acidic soils do best at this balance. Notice that the sum of calcium and magnesium saturation is 80%. Ideally, you want 80% of an acidic soil's TCEC saturated with these two elements.

Many organic gardeners have over-used dolomite lime. A soil test will reveal if over-use has produced large magnesium excesses. If this is your situation, it can be fixed, but don't expect to achieve 68:12 in one year. Expect flocculation to happen gradually; expect to slowly taper off your extreme use of organic matter, not to slash it immediately.

When the ratio goes the other way — when calcium is excessive, but magnesium is deficient — the soil can become extremely loose. If water seems to flow through your soil without sticking, *and if the magnesium saturation is below target,* this will get better as the magnesium comes up. Otherwise, be grateful it isn't tight and airless.

The Worksheets

We have been through this once before, but it's so important I'm going to say it again. Suppose a soil has an TCEC of 1. This means it has a precise (and huge!) number of attachment points in the top six inches of an acre. If a total cation exchange capacity of 1 were to be totally saturated with calcium cations, such that the *only* cations this layer of soil held onto were calcium cations, then the total weight of calcium involved would be 400 pounds (refer again to Table 5.2). If we completely saturated the furrowslice acre of a soil with a TCEC of 1 millequivalent with *sodium* cations, there would be 460 pounds of sodium present. If soil with a TCEC of 1 meq were to be holding one-half a meq of calcium and one-half a meq of sodium, it would have 200 pounds of calcium and 230 pounds of sodium. For that situation, we could also say that the soil has a *base cation saturation ratio* of 50% calcium to 50% sodium.

The top six inches of Matthew's soil has a TCEC of 13. If it were 100% saturated with calcium it would hold 400 x 13 = 5,200 pounds of calcium. But that much calcium would leave no room for any other nutrient. In the real world, total calcium saturation only happens in a test tube. And, anyway, for most soils, we want 68% calcium saturation. So the equation we want is: 400 x 12.87 x 0.68 = 3,501 pounds of calcium. And we write that (rounded-off) number into the space on the worksheet for "Target."

To work out the target weight of magnesium, multiply the weight of 1 meq of magnesium by the TCEC, and multiply that by the percentage saturation desired. Or: 240 x 13 = 3,120 x 0.12 = 374 pounds. Write that into the space on the line for the magnesium "Target."

One advantage to using Logan Labs is that their report form is based on 68:12, so their desired levels and any deficits for calcium and magnesium will be exactly as you'd calculate them yourself.

Calcium **Ca**	lb/ac 3292	**TCEC x 400 x 0.68 = Target Level**	Minimum target 1,900 lb/ac 3536	244
Magnesium **Mg**	lb/ac 210	**TCEC x 240 x 0.12 = Target Level**	374	164

Fig. 5.4.

Potassium

Plants concentrate potassium into structure — stalks, stems and fiber. With grasses and cereals, potassium will be uniformly distributed throughout the plant until seed starts forming. When flowering begins, the most valuable nutritional elements like phosphorus, nitrate, sulfate, etc., are translocated out of the no-longer-growing leaf and stem cells and placed inside the seed coat, where they are put into storage around the embryo to provide it with a full and balanced nutritional storehouse to use during sprouting. Not so with potassium. Locked tight in plant structure, it remains in place. Thus, hay and straw contain a lot of potassium. Trees and shrubs concentrate potassium in their woody parts; most of the other nutrients move into the leaves, where they mainly form chlorophyll, and/or flowers, fruits and seeds. Thus, sawdust and bark are potassium-rich.

The point: gardeners who import masses of organic matter to make compost or to use as mulch usually import the waste products of grass agriculture. (And sometimes, unfortunately, forest wastes.) They use spoiled hay and cereal grain straw and lawn clippings. They bring in manure that comes from grass-eaters, like horses, cows and sheep. If they use fallen tree leaves, these have already returned most their valuable minerals to the tree's sap for storage over winter. So, when a gardener sets out to build soil fertility through the importation of massive quantities of decomposable organic matter, they usually import a lot of potassium and comparatively less of the other plant nutrients. Soils handled this way do not produce nutrient-dense food.

This book targets even lower potassium levels than most biological farming advisors call for. That's because all farmers, including those using organic or biological methods, are dominated by economics. Farming these days, is, by definition, a business — not a practice of self-sufficiency or a hobby that earns a bit of pocket money on the side. Farmers of all persuasions must make a profit, if only to maintain the illusion that they own land instead of having the right to occupy it indefinitely (but not absolutely), so long as they continue paying the state an annual rent. Makes me wonder if this underlying reality — the state really owns the land — inclines farmers to a short-term approach. Whatever deep social currents may be the real cause, today's farmers try to produce the highest possible yields at the lowest possible

Sufficient: This is an agricultural term referring to targeting nutrient levels that provoke maximum yield with minimum input. This book calls for levels considerably higher than the usual farm consultant would consider sufficient (or affordable). However, concerning nitrogen and potassium, I recommend lower levels than are usually considered "sufficient."

cost of inputs. And the most inexpensive and effective yield-booster is potassium.

Carbohydrates and fiber are constructed from potassium, carbon and hydrogen. Plants get plenty of carbon from the carbon dioxide in the atmosphere. Hydrogen — no shortage of that, every drop of water in the soil contains hydrogen. So, if potassium is abundant, and there is sunlight and moisture, the plant makes an abundance of carbohydrates, sugars, fats and fiber. But to make proteins, enzymes and vitamins — the important stuff — plants need the other, often scarcer elements: nitrogen, phosphorus, sulfur, zinc, copper, iron, manganese, magnesium, etc. If these elements are not critically scarce, plants can still produce carbohydrates. When they aren't scarce, plants make valuable nutrition in balance with their carbohydrates. And when potassium is just a little bit scarce, plants make the highest concentration of nutrition, which is what we need to eat in order to be healthy.

William Albrecht pointed out a key regional difference in North American soils: the less-leached ones have a lower percentage of potassium saturation, while the well-watered soils that once grew forests offer plants a greater potassium saturation. Remember, saturation percentage is not quantity, it is about balance: an unleached soil will offer more potassium overall than a leached soil does, but relative to that abundant potassium, there will be a matching, and greater, abundance of the other essential elements. A leached soil, on the other hand, will still offer a lot of potassium, but it provides a relative scarcity of the other nutrients. Albrecht analyzed crop quality, region by region, with the evapotranspiration ratio map in mind; he discovered that the foods from less-leached soils are far more nutrient-dense, and contain much higher levels of protein, phosphorus, calcium and magnesium, but offer relatively less potassium. Foods from well-rained-upon soils provide the consumer with relatively more potassium, fiber and calories, but correspondingly less protein, phosphorus, calcium and magnesium.

To appreciate the health consequences arising from excess soil potassium, consider choosing between two imaginary batches of potatoes that you're going to depend upon as the family's staff of life. Imagine it is really hard times, and your family is going to subsist on spuds, like millions of oppressed Irish cottagers were forced to do in the early 1800s. The first batch is industrial potatoes, grown using the latest and best agronomy; the second is a homestead batch, grown so as to be the most nutrient-dense possible. The homestead spuds receive mega nutrition in all respects *except for potassium.* The industrial potatoes get an agronomic sufficiency of K, and consequently yield 25% more tons per acre containing 25% more starch by weight (for starch, read: more calories), but 25% less protein and 25% fewer minerals and vitamins by weight compared to nutrient-dense potatoes. Thus, the industrial potato farmer produces adequate number of bushels to profit enough to keep on growing spuds, but the poor human bodies trying to survive on those tasteless potatoes are driven by hidden hungers to overeat in search of proteins, vitamins, minerals and vital enzymes. The consequence is disease of all sorts. Now, if you please, imagine that it is present time, and *all* the foods in the supermarket are similar to those nutrient-undense spuds I just described. And imagine that the diseases consequent to the hidden hungers induced by those imitation foods are falling on you and yours.

Heavy soils deliver potassium far more effectively to plants than light soils do. To produce nutrient-density, heavy soils should be brought to a lower percentage potassium saturation than light soils require. The smartest agronomists I know of target potassium at 5% or even 6% of the TCEC on very light ground, down to 2% saturation — or even less than that — on extremely heavy soils.

Those who practice balancing pay close attention to element associations. Iron and manganese have a relationship; phosphorus and zinc do too. For the best result, these elements should be present in a broad range of ratios (for example, iron should exceed manganese by at least one third). Often, those ratios hold good through the full range of soils; light, medium and heavy all require the same proportions, though in larger quantities as the soil becomes heavier — but still in the same proportions. One widely accepted ratio is P = K. Gary Zimmer, a well-known soil advisor and author, says ideally Zn = $^1/_{10}$ P because if either one gets far from this ratio, it interferes with the uptake of the other — excessive zinc can induce a phosphorus deficiency while excess phosphorus can induce a zinc deficiency. My own use of Cu =

Potassium Target Levels					
TCEC	**Pounds**	**TCEC**	**Pounds**	**TCEC**	**Pounds**
If TCEC Is below 7,		16	390	28	493
use the value for 7		17	400	29	500
		18	410	30	507
7	255	19	420	31	511
8	270	20	435	32	515
9	290	21	443	33	519
10	310	22	451	34	523
11	320	23	459	35	527
12	335	24	463	36	531
13	350	25	475	37	535
14	365	26	481	38	539
15	380	27	487	39	543

If TCEC is above 39, use the value for 39.

Table 5.3: *Potassium (K) Target Levels.*

½ Zn is a useful convenience for quickly coming up with a desirable copper concentration, but I make no assertions that zinc and copper have any interactions.

In my opinion, potassium saturation should shift with the exchange capacity, that is, with the ability of the soil to deliver potassium. Unlike the rest of the elements, no single ratio can be used to work out the right quantity. Instead, make it simple: use Table 5.3, shown in the sidebar. It also appears on the worksheet. When filling in the worksheet, make K = the amount on the table that matches the soil's exchange capacity. Transfer that number to the worksheet in the "Target" column.

Sodium (Na)

Calculate the sodium saturation target in the same way you did magnesium and calcium (see sidebar).

People are often shocked at the suggestion that they should spread sea salt on their gardens. Most of us learned how the Romans made

Sodium

1 meq sodium weighs 460 pounds. If we have a soil with a TCEC of 13, calculating 2% saturation with sodium would be done this way:

460 x 13 (the TCEC) = 5,980 pounds of sodium. This amount could entirely saturate the clay fraction of a 13 TCEC soil. And 2% of 5,980 pounds is 120 pounds; that amount constitutes a 2% saturation at that TCEC.

sure that the Carthaginians could never recover from being defeated by strewing their fields with salt. It is true; too much sodium wrecks soil. But not at 2% saturation. And some crop species need quite a bit of sodium.

Recall that calcium causes clay to loosen, and magnesium makes it tighten. Sodium makes it tighten, too — and a lot more powerfully than magnesium does. A soil holding excess sodium shrinks to an airless condition in which plants do not grow well even if they can tolerate the saline conditions. In some regions of North America, particularly the arid areas, soils naturally contain a lot of sodium. And even more regions have significant levels of sodium in their water supplies, so sodium gets deposited into a garden when it is irrigated. Knowing what you know now about pounds per acre and parts per million, suppose that your municipal water system provides water with 50 ppm sodium in it. That number, 50, was not chosen at random. Erica Reinheimer gardens around Arroyo Grande, California; her municipal supply has that average sodium level. Fifty ppm sodium in an acre of soil six-inches deep weighs 100 pounds. Erica spreads more than two acre-feet of water during the summer growing season. If a six-inch deep layer of water has 100 pounds of sodium in it, then a two-foot-thick layer of irrigation water would supply about 400 pounds. Add that much sodium to your soil for a few years running, and you're asking for big trouble unless there are heavy winter rains to leach the soil or the gardener takes steps to chemically remove the sodium (with gypsum). Sodium is only readily leached from soils that are well-saturated with calcium, magnesium and potassium.

The sodium levels in municipal water vary considerably. As just mentioned, in Arroyo Grande, California, Erica's water has 50 ppm sodium; Portland, Oregon's high quality water supply still contains 3.5 ppm — which brings with it 75 pounds of sea salt per acre per year if you provide two feet of irrigation during the summer.

The *Acid Soil Worksheet* specifies 2% saturation, but I suggest setting your own sodium saturation target using the following as your general strategy: If there is any risk of having more than 5 parts per million of sodium in your irrigation water, target 1%. Keep the sodium target at 1% if you garden in semi-arid or arid soils; and keep in mind that irrigation water in these areas commonly has sodium in solution. However, if your garden is normally well watered by rainfall to the point of leaching it once or thrice a year, and if you are confident of the purity of your irrigation water, then up the sodium target saturation level to 2%. Anyone depending on irrigation would be well advised to have their water tested for sodium (and other contaminants). If you are using municipal water, there should be an analysis available for the asking.

There are ways to reduce excess sodium. They will be discussed in the section to come dealing with handling excesses.

Summary

If all we knew how to do was harmonize the four major cations and make compost, our agriculture and gardening would be enormously improved. We could conduct this balancing act without risk of ever running out of raw materials. Calcium and magnesium come from limestone (or gypsum); we merely have to grind this commonly found soft rock finely and spread it. The real trick is to get calcium and magnesium into the right balance and quantity. Sodium, we get from sea salt. No problem; lots of that. Potassium? Well, there's a potassium-rich rock called greensand (sometimes called "Jersey greensand" because it is found in New Jersey). There are many common rocks that have high potassium levels. Another source is wood ash. So, even in rather primitive conditions (if we somehow could run a soil lab), we could bring these minerals into balance on our farm soils. In most cases, knowing what we know now, we could also balance the major cations without ever testing the soil. If we only did the following, we'd grow enormously better food than we do at present: spread enough ag lime to take the actual sour taste out of the soil, plus only enough dolomite that the soil does not tighten up, spread finely crushed, high-potassium granite about half as thick and a quarter as often as the ag lime; and annually supply about 50 pounds of real sea salt per acre on most soils.

Fortunately, we *do* know how to go beyond balancing the four major cations. As each aspect of plant nutrition is brought into balanced abundance, the result improves.

Computing the Percentage of an Element

A fertilizer bag that says its contents include 5% nitrogen as nitrate, NO_3, does *not* contain 5% nitrogen. Perhaps that labeling convention was designed to mislead. In any event, here's how to calculate the truth of the matter. The atomic weight of one atom of nitrogen is 14 and a little bit. (That number, 14, is not exactly right; the atomic weights I provide are rounded off, making them easier to work with.) And the atomic weight of one atom of oxygen is about 16. So in NO_3 there are three atoms of oxygen and one of nitrogen, the percentage of nitrogen is computed this way:

3 x atomic weight of oxygen = 3 x 16 = 48

plus atomic weight of nitrogen (14) gives:

the atomic weight of NO_3: 48 + 14 = 62.

The atomic weight of N divided by the atomic weight of NO_3 gives the percentage of elemental nitrogen in nitrate: $^{14}/_{62}$= 22.5%.

Calculating soil remineralization is more straightforward if you use only the elemental weights.

By the way, the fertilizer industry does the same fiddle with phosphate and potassium (oxide). Elemental phosphorus is labeled as phosphate, P_2O_5, which is only 44% elemental P, and potassium is normally labeled as K_2O, which is only 83% potassium.

To make your life easier, the worksheets contain a table of fertilizers showing their contents in elemental form. When the table says seedmeal contains 6% N, it means elemental N. If that 6% N is in nitrate form, NO_3, you would figure 6 divided by .22, giving you 27% nitrate nitrogen.

The Anions

Anions attach to humus, but not to clay. If soil lacks organic matter, the crop usually suffers wildly shifting nutrition levels. If you added phosphorus the previous year, and little or nothing of it shows on the next spring's soil test report, it's likely your land could use more humus and/or better-quality humus. Phosphate that fails to get into the organic fraction soon goes insoluble and also disappears from next year's test score.

I look at the apparent waste of anions this way: when I take B vitamins (or vitamin C), my kidneys end up removing it from my blood, and I end up with yellow urine. An M.D. would sneer and say I was just pissing away money. It could seem logical to assert something like that about wasting sulfur or phosphorus if it fails to appear on the next soil audit. On the other hand, nutritional healers understand that vitamins help far more at high blood concentrations, and that the body is *supposed* to eliminate them from the blood, and you're *supposed* to constantly be ingesting them. This analogy doesn't quite work with the soil, but suppose that anions at high concentration do something to nutritional outcomes; suppose, even if they are "wasted," they still did a lot of good.

Phosphorus (P)

Phosphorus fertilizer seems expensive, especially when building levels that create nutrient-density. But, pushing soil phosphorus levels well beyond sufficiency seems highly desirable to me, despite the cost. Certainly you do not want low phosphorus! Phosphorus determines the speed at which plants grow because it is a key part of all cellular enzymes, including those that liberate and transfer cellular energy. I like this analogy: If you lower the voltage (less P), the motor doesn't spin as rapidly. The most confusing thing about phosphorus nutrition is that when plants are short P, they usually manifest no obvious symptoms other than slower growth (something home gardeners can rarely gauge), leading to a smaller ultimate yield of lower nutritional quality and poorer flavor. Yes, when phosphorus is catastrophically short, the plant may turn purple and be obviously distressed, but even slight deficiencies immediately reduce growth rates. Of this, Carey

Reams (1903–1985), who doctored soils and wrote books about it, said:

> *The factor which determines the mineral content in any produce, whether it is a grass, or anything else, is the phosphate in the soil. The higher the water-soluble phosphate, the higher the mineral content. In order to get the maximum amount of nutrient in the crop, and the maximum yield, a minimum of 400 lbs. per acre of available phosphate is needed. That much cannot be supplied from superphosphate, triple superphosphate, or hard rock phosphate. Soft rock phosphate is the best way to achieve this level, besides its having many other benefits.*
>
> (source: customers.hbci.com/~cmills/PHOSPHATE%20Reams.html, accessed July 7, 2012)

I point out Reams said "available *phosphate*," not available phosphorus. And I remind you that P = 0.044 x P_2O_5. One hundred seventy-five pounds of *elemental* phosphorus per acre is a gracious plenty. With enough organic matter present, it is possible to usefully apply 175 pounds of available P on very light soils. Really heavy soils can, and should, hold up to 500 and some pounds of P.

I've been stressing the importance of phosphorus because this is one nutrient people are tempted to cut corners on. Phosphorus is costly when supplied at luxury levels, and plants seem to do fine with far less of it than what it takes to produce real nutrient-density. Phosphorus is increasingly scarce. The planet is experiencing peak phosphorus in much the same way we have passed the point of peak oil production. The price is inevitably going to go up, and then it is going to go up some more. Despite the cost involved, I urge you to bring your gardens to the highest useful level.

Few farmers fully remineralize phosphorus, or any other element for that matter. In fact, farmers have been mostly running phosphate mines and calling them grain fields. Farmers often supply only the amount of P sufficient for the crop being grown. Gardeners, on the other hand, are not dealing with acres by the hundreds or thousands;

they are dealing with a few hundred or a few thousand square feet. And what if it does cost us a few dollars more to produce our food? Compared to the supermarket price, no matter what the cost of our inputs, home-garden produce still comes out far cheaper in dollar terms. If nutrient-density is considered, comparing the monetary cost of production means nothing.

This year, I purchased a ton of Queensland soft rock phosphate (SRP) in sacks. The price was about A$800 in Queensland, plus freight to Tasmania. (At the time I am writing this, the Aussie dollar is about par with the US dollar.) Queensland SRP assays at 8.8% elemental phosphorus, so one ton contains 176 pounds of elemental phosphorus. Forget the cost of freight. Forget that it was a metric ton, so it's weight actually was 2,200 pounds. If this had been a farm-scale purchase, a full shipping container of unbagged SRP could have cost me less than $600/ton, delivered. This year. Next year, it probably will cost more.

A century ago, the furrowslice acre of an average North American farm, in an area where the evapotranspiration ratio was over 100, held 2,000–5,000 pounds of elemental phosphorus. Once. Originally, the biological processes in that average farm soil released more than sufficient phosphate during the growing season. But soil erosion and crop removals have taken away so much of that original endowment and so weakened the soil's microlife that now the farmer must use phosphate fertilizer with every crop. An acre of grain sends 20–40 pounds of elemental P through the farm gate. If a farm was put into production 100 years ago, and 20 lbs/P/ac/year has been taken away ever since, then over the past century something like 2,000 pounds of P per acre have mostly gone down various outhouses and sewerage systems (unless the grain was fed to farm animals). It might seem that removing 2,000 pounds of phosphorus from a soil that held little more than that amount to begin with might not leave enough to allow it to continue to produce crops. Keep in mind that I was referring only to what was originally present in the average topsoil, not to what accessible treasures might lie below that level.

The average farmer in this scenario probably applied some phosphorus fertilizer in the past decades, but the cropping system lowered the soil's organic matter levels. This hugely reduced the soil's anion exchange capacity, and down went the phosphorus level. I consider the removed fertility as an off-balance-sheet national debt owed until repaid — hopefully not with too much karmic interest attached. And what would the dollar price be for the 2,000 pounds of P, not counting the other minerals that also departed? Well, in the form of bagged soft rock phosphate delivered to Steve Solomon's property, the bill at this year's price would come to about $10,000. Per acre. And that's just for the P that has been extracted.

I have just demonstrated why a deeply debt-enslaved farmer cannot give up soil mining. Nor could a free-and-clear farmer consider full remineralization because they, too, must sell to a market in which the price is determined by soil miners. But if you are intending to remineralize a really big veggie garden, say an eighth-acre garden, then instead of confronting a $10,000 per acre phosphorus debt, you're looking at $1,200. But the debt of all that lost phosphorus could not possibly be repaid at one go, even if you wanted to. If you put in a gracious plenty of phosphorus with every crop and build your soil organic matter, then the soil background levels will gradually build back up until they reach target for that TCEC. Then you can stop adding phosphorus, probably for a decade or two.

A classic book about the economics of farm remineralization is *The Story of the Soil* by Cyril Hopkins (1903). It is about an intelligent young man looking to buy an exhausted farm in Virginia on the cheap. He fully considers the cost of remineralizing that land using rock phosphate and lime. Hopkins knew the whole story a century ago. You can download the book for free; full details are in the Bibliography.

So what is the best overall strategy to build P? First, increase the anion exchange capacity. If the phosphorus you add gets hooked up with humus or comes already as a part of manure or compost, it will remain available for a long time — maybe 20 years, maybe 100 years. But if P fails to connect with humus, then it will almost inevitably hook up with calcium or worse, with iron. It will become one of the insoluble forms of calcium phosphate, or the extremely insoluble iron phosphate. And if this happens, your expensive P will invisibly merge into the soil's background reserve. Only a tiny fraction of that reserve will be available on next year's soil audit. Once the reserve is large enough, there will be plenty of available phosphorus.

Someday, you may be able to reduce phosphorus additions to the small amounts crops actually remove each year. However, most soil advisors say when you add 10 pounds of available phosphate to a soil that is very low in phosphate, you'll be lucky to find 1 pound added to the level on next year's test report. But the relationship improves over time. As the soil becomes more saturated, a larger proportion of applied phosphorus sticks. However, take heart: those soil advisors are not dealing with garden soils carrying 5% or 7% organic matter.

Phosphorus

Target Level is P = K.

Single applications are limited to 175 pounds per acre.

Matthew's target is 350 pounds per acre. He has excess P.

Without a limit of 175 pounds per acre per single addition, complying with P = K could prove extremely wasteful. And for sure, 175 pounds is generous. It makes sense to spread P with generosity because ultimately you will use less fertilizer than if you gave the soil just a little more than the current crop needs. Suppose your target level is 335 lbs in the top six inches of an acre (you're working with a soil having a TCEC of 12.0). You have the type of ordinary, maybe average soil, often seen. Suppose that soil has almost no available phosphorus, another thing often seen. If the cost of elemental P is US$5 a pound, then 335 pounds of P costs about US$1,600 as soft rock phosphate, which would mean about US$200 for an eighth-acre veggie plot.

If money is abundant, you could start by adding 335 lb/ac, and then, as the starting level came up, you could feed slightly less phosphorus every year until eventually the spring soil audit approached 335 pounds. Actually, our hypothetical soil would inevitably be quite low in organic matter as well as phosphate, but since this is all imaginary anyway, suppose it had quite high organic matter level, just no P. In that case, much of your expensive 335 pounds of P might stick. Vegetable crops growing on light soil that is well below phosphorus targets will do quite well if their soil is fed even an additional 100 lb/ac of actual phosphorus (230 lb/ac phosphate). However, when you retest that soil a year later, the background levels may not have increased by much at all. But if you'll apply a generous 175 lb/ac to that soil, you'll get a great crop and see some meaningful build-up next spring.

You can marry phosphate fertilizers into compost and thereby make a far larger percentage of composted rock phosphate become available. Already part of organic matter the phosphorus remains available so your levels will build up faster. This will be discussed further in Chapter 9.

Sulfur (S)

Sulfur, in partnership with nitrogen, forms key pieces in several essential amino acids and crucial enzymes. When sulfur is abundant, these are plentiful; in consequence, the plant is able to form a broader range of proteins. Flavor is more abundant and so is nutrient-density. When sulfur is short, plant proteins are less complete and have lower feeding value. Sulfur has long been used as a fungicide; ground to fine powder, it is dusted on plants to prevent or fend off diseases. I suspect much of the reason sulfur works to fight disease is because the diseased plants were seriously short S in the first place.

A few decades ago, most of North America received an ongoing acid-rain sulfur hit from the burning of coal and other fossil fuels. Now, acid rain has largely been cleaned up by environmental legislation, so farmers have to pay attention to getting sulfur into their fields. So must you.

Putting high levels of S into topsoil leaches out cations. This leaching is not necessarily something to be avoided. When sulfur merges into the soil solution, it takes the form of the sulfate anion, SO_4^{-2}.

Sulfur

Target S = ½ Mg until there are no more cation excesses. Then:

S = ⅓ P.

Or, if you want to improve your subsoil, S = ½ Mg.

When you till in raw sulfur, which is an excellent fertilizer, albeit a bit harsh, it is converted to sulfate ions by soil bacteria. Every combination with sulfate I know of is highly soluble in water: iron sulfate, zinc, copper and manganese sulfates, potassium sulfate, etc. Even uranium sulfate is water soluble. Calcium sulfate (gypsum) is also soluble, but not quite so readily as the others. Should too much rain or irrigation flow through the soil, it can leach sulfates. Thus, having high sulfur levels results in a steady (slow) reduction of nutrient levels in the topsoil. Farmers, counting the bottom line, have to work hard to keep costly fertilizers in the topsoil, so they might not want more S present than what is sufficient for the crop. The sulfate anion's ability to connect with cations and keep them in soluble form is why this book specifies a high level of sulfur when there are excess cations to leach out.

SUBSOILS

Slow leaching can provide the garden a huge benefit. If you have a subsoil that is capable of holding onto cations, then what you leach out of the topsoil can be captured and retained there. If what leaches down is balanced nutrition, then your subsoil will be enormously improved thereby. Subsoils in the eastern United States and Canada are usually more acidic than their topsoil is. Most are nearly pure clay that originally formed in the topsoil and then was transported into the subsoil by water and deposited there. Subsoils tend naturally to be airless because they are far from being saturated by calcium; their clays are packed so tightly that no roots could breathe in them — even if there weren't toxic levels of manganese and/or aluminum, which is usually the case. Skip ahead and look at Figure 5.5. It is a chart that saves me a thousand words of description and deserves several meditations on your part. It shows how nutrient availability changes with soil pH. Manganese, an important nutrient, becomes 100 times more available at pH 5.0 than it is at 6.0 (and it is also toxic at those concentrations). An even worse problem under acidic conditions is the increased availability of aluminum. Even extremely low levels of soluble aluminum are highly poisonous to plants. Fortunately, aluminum is almost entirely insoluble at pH levels over 5.0. But few plants can tolerate the amount of aluminum available when pH is below 5.0.

Gardeners imagine that vegetables root only in the topsoil; this is not true. An agricultural scientist named John Weaver found that most species form root systems at least four and even six feet deep on Nebraska prairie soils (which have relatively open, non-acidic, free-draining subsoils, well supplied with air). When the same varieties were grown in New York State, where there was an acidic clay subsoil, they made root systems only about two feet deep. (If you're interested in learning more, Weaver's classic book is *Root Development of Vegetable Crops* is available for free download at soilandhealth.org.) But, I'd bet that if Weaver had amended his Nebraska land with gypsum at a ton or two per acre for a few years prior to growing those vegetables, his team would have found roots going even deeper, and they would have been far more densely developed. It won't be too many more pages before you fully understand why gypsum will cause that sort of transformation. And I'll give you a theory: That field in New York that grew a few of Weaver's vegetables with restricted root systems...I bet that before the original old-growth forest had been cleared, the subsoil was not impervious to root penetration. For the first years that field was used for farming, its subsoil remained open to root penetration, which is why yields were so large in those early days, and the food grown was nutrient-dense. But the subsoil was gradually robbed of minerals, until it became too acidic for crops to root into it any longer.

Tiedjens tells us of another way subsoils were wrecked by farming; this type won't have an acid pH, even if they are in a climate that normally has acidic soils. When potassium chloride (KCl) is heavily used as fertilizer (which it was and still is), the intense flush of potassium being released knocks some calcium cations off the exchange points; these combine with the chloride, forming calcium chloride ($CaCl_2$), a highly soluble substance that is easily leached. In short, for every unit of KCl spread on the topsoil, a goodly amount of $CaCl_2$ leaches into the groundwater. As the subsoil becomes depleted of calcium, its exchange points get filled with potassium. Thus saturated, the subsoil is not acidic; in fact, since potassium raises pH more than calcium does, it will be slightly alkaline. But it will contain so little calcium that roots cannot grow there. For all intents, that subsoil is dead.

Remineralizing the subsoil by digging lime and other fertilizers directly into it would be an exhausting task if done with a spade and wheelbarrow. I've always wondered why those intensive organic gardening books so glibly recommend putting in SaMOA two feet deep. Well, just you try to double-dig your way through a gooey clay subsoil. Just try! But there's an easy solution. Allow elemental sulfur in the topsoil to hook up with a good bit of calcium if there's an excess of it, or else add gypsum — calcium sulfate — and cause those elements to naturally leach down into that acidic clay. As the subsoil gradually becomes filled with cations, especially with calcium, its acidity will moderate, or its potassium excess may be reduced, which may allow root penetration.

When setting food-garden sulfur targets, the fundamental questions are: Are you planning on using that land long term? Do you want to remineralize only the topsoil or remineralize the topsoil and eventually, gradually, the subsoil? You'll still get a decent, nutrient-dense crop without bothering about subsoil. But you'll get far better long-term results if you can improve subsoil conditions. The deeper the roots can go, the larger and healthier the plant will grow. Plants with accessible subsoil can grow larger than most gardeners think is possible. Plants on deep, open soil get enough elbow room to achieve extraordinary size without getting stressed by having to compete too much. This means fruit-bearing crops like pole beans, cucumbers, zucchini and tomato do not necessarily have their yield taper off after an initial burst; instead, the yield keeps on increasing as the plant keeps on growing. This approach is the opposite of intensive gardening, and it's something I've encouraged for many a decade. An open subsoil also makes the garden far more drought-proof because there normally are huge reserves of moisture in subsoil clay — if only the plants can establish roots in it.

And wouldn't it be fine to leave your land more productive than you found it?

So why not dig yourself a three-foot-deep hole in your garden and make the acquaintance of your subsoil, if you have one. This is most easily done with a soil augur of the sort used to set fence posts. To get three feet down, you may need to extend the augur's shaft; this isn't difficult. Otherwise, believe me, the result is worth an hour of pick-and-shovel work. The subsoil may look like gooey, airless stuff with no sign of root penetration anywhere, but it might be fixable. And what if, to your delight, you find your subsoil ain't that bad, and you find some grass roots a yard deep? In either case, be a big spender; run two standard 20-buck soil tests using a 12-inch sample depth. Take one sample in the top foot of soil; the other sample in the next foot down. If you manage to dig as far as the third foot, be a bigger spender and sample the third foot, too. If you're not in the humid Southeast of the United States, where most of the clays are old and tired, and if you are east of the 98th Meridian, you might be surprised to find your subsoil clay has an astonishingly high TCEC. Like 50 or 60. If you're over 100 on the evapotranspiration map and not in limestone country, it will be acidic — around pH 4.5. The aluminum level may be way too high. And there may be toxic levels of manganese, too. Now, take another look at Figure 5.5 and imagine what that clay could be if you managed to even partially saturate and somewhat balance its huge TCEC and thereby get the pH up to a comfortable range.

If you live where the evapotranspiration ratio is under 100 but over 60, you may not have a clay subsoil at all. Instead, the same stuff making up the topsoil will be found in the subsoil too, although the percentage of clay in it will be higher than it is in the topsoil. I expect the soil in which John Weaver's students exposed the finest details of plant root development by gently toothbrushing away the soil was of that sort. These kinds of subsoils are worthy of great investment, especially when all you have to do is to remove some of that magnesium and/or potassium, replace it with calcium, and thereby open up another foot or two of soil to root penetration. For my retirement, the Universe seems to have gifted me with a subsoil like that.

Suppose you are living on a glacial moraine, such as the sand hills of western Washington State or the Kootenays in British Columbia (two similar locales I know well from having lived there). As you dig four feet or even ten feet into your coarse sandy soil, you'd find only more coarse sand and rounded glacial rocks. There is nothing down there capable of holding cations, so encouraging valuable plant nutrients to leach into that subsoil would be pointless. Your evaluation may be quick and easy if your garden has only a foot or two of topsoil, with solid rock below. In that case, there's no sense having valuable plant nutrients sliding along that interface between soil and rock, heading downhill on a subterranean journey to the ocean.

There's another reason to get into communication with your subsoil. If you happen to have a sandy garden with a clay subsoil, and that clay just happens to have a high cation exchange capacity (20 to 80), then you will be able to make far better compost if you add one or two percent (by starting volume) of that high-cation-exchange-capacity clay to it. Clay like that is almost entirely missing from your topsoil; it's already been transported into the subsoil. And such clay won't be found for sale on the shelves of garden centers in a convenient dry powder form. But if you have some good clay already on hand, if all you have to do is to dig a deep hole and there it is, a bit of clay mining is worth considering. And speaking of being a law-abiding citizen, did you ever see exposures of clay along minor highways?

Boron (B)

Two pounds of boron in the furrowslice acre (1 ppm) is the minimum concentration at which no obvious deficiencies appear in most soils. Three pounds per acre is considered sufficient by most agronomists. More than four pounds of boron (2 ppm) in the same volume of soil can be toxic to a few crop species, but is comfortable for most. My main goal regarding boron levels is for everyone to have a good result — and be safe at the same time. At worst, mild excesses will leach out in a few years. I can tell you this: If there is boron toxicity, the first crop to show it will be green beans. Celery, potato, tomato, radish, corn, pumpkin, peppers (and chilies), sweet potato and lima beans are in the "semi-tolerant," group, meaning they can take only so much boron, and then they get in trouble.

Because fertilizers can so easily be double applied (or more than that), this book specifies boron applications be limited to two pounds per acre. If the soil called for that much it would be possible to put in another two pounds of boron one month after the first amendment, but generally, two pounds of boron in one year is more than sufficient to feed soil and hopefully increase the boron level next year. Even though boron is applied in minute quantities, it is easy enough to uniformly spread boron all by itself because borax is readily soluble. If you dissolved 20 grams (roughly one heaping tablespoon) of laundry borax into one quart of hot water, and sprayed that water fairly uniformly *on the soil* of a 100-square-foot bed, you'd be applying 2 pounds per acre actual boron. That concentration will not damage leaves, but to be effective, boron must go into the soil. It is poorly used by plants when sprayed as foliar fertilizer.

Target Level for Boron

Light soil, TCEC below 10.0 = 2 lb/ac (1 ppm)

Heavy soil, TCEC above 10.0 = 4 lb/ac (2 ppm).

Even though it's powerful stuff for the amount needed, please don't be scared of boron. It has a unique and vitally important job to do. The minuscule tubes that plants use to conduct moisture must be lined with boron. If these tubes don't get enough boron, they don't function effectively, which means the plant can't drink effectively, which means it can't nourish itself effectively. As Hugh Lovel, a biodynamic soil consultant puts it: boron comes first, and then come the rest of the plant nutrients.

A plant's vascular tubes are also the lodging point for much of the plant's acquisition of silicon. Boron and silicon work together at an atomic level to move moisture up these minute tubes. Silicon, yet another anion, is one vital plant nutrient that is not yet appreciated by most soil analysts and so is not routinely tested for, and there are few soil amendments sold specifically to raise silicon supplies in soil. However, soft rock phosphate usually contains a good deal of silicon. Silicon is one element where SaMOA won't hurtcha.

Nitrogen (N)

Nutrient nitrogen in the soil will be in one of two chemical forms: nitrate (NO_3), an anion, or ammonium (NH_4), a cation. A standard M3 soil test does not report the level of either of these. That's because conventional farming runs soil organic matter levels way down and then uses chemical nitrogen to compensate. In consequence, nitrogen levels are unstable. They move up and down rapidly with the season and with the crop cycle; knowing what level of nitrate or ammonium was available a few weeks ago doesn't help a farmer a great deal.

However, without testing for it specifically, there is a way to anticipate how much nitrogen will be usefully released by a garden soil. Even relatively new gardens can, by themselves, provide enough nitrogen for low-demand vegetables; maybe even do better than that. The source of this nitrogen is the ongoing decomposition of soil organic matter. The amount of nitrogen released depends on how active the biological systems are (determined by the soil-air supply and nutrient balance) and on the existing level and quality of soil organic matter. Finally, the rate of nitrogen release changes greatly with the soil's temperature. In the cooler parts of North America, a common hurdle involves getting the spring nitrogen level up high enough that early crops can get growing fast.

Gardeners do not need to predict nitrogen release with precision. It is workable to assume that the standard equation is correct: the annual quantity of N (released) = 15–25 pounds actual N per acre per 1% of soil organic matter. I think it best practice to anticipate the lowest possible level of nitrogen release. If your soil contains 5% organic matter, assume you'll get 75 pounds of actual nitrogen per acre released over the summer. Unfortunately, most of that natural nitrogen appears during the warmest two months. For this reason, adding organic nitrate fertilizer in spring is essential in many regions if you want to get spring-sown crops to produce well. For example, Cascadian soils warm up slowly at best; in Oregon gardens, I've seen many instances of nitrogen-deprived sweet corn during June. It usually turns properly green in July, but, because it was deprived during June, the corn isn't knee high by the 4th, and in consequence, yields much less. That is precisely why COF is so popular in Cascadia; it introduced the region's compost gardeners to the benefits from using concentrated nitrogen fertilizer.

Nitrogen is the key element required to form proteins, and protein is the very stuff of life itself. All other factors being in a reasonable range, the amount of nitrogen in the soil controls how much soil-protein (which effectively means how much soil ecology) you're going to have working for you, because microorganisms are basically little bits of protein that eat soil organic matter. Thus, adding nitrogen increases the speed at which soil organic matter is going to disappear. By keeping soil nitrogen levels to the minimum needed, you stop unnecessary loss of soil organic matter.

Nitrogen

Annual nitrogen release from soil = 15–25 pounds actual N per 1% soil organic matter.

To convert nitrate (NO_3, the analysis on a fertilizer bag) to elemental N, use this formula:

$$N = 0.22 \times NO_3.$$

Typical of business (un)ethics, a load of fertilizer that is labeled at 100 pounds of nitrate nitrogen actually contains only 22 pounds of nitrogen. Buyer, be aware.

Protein, on average, contains 16% N. So a 50-pound bag of cottonseed meal, at 45% protein, contains 3.6 pounds N. The calculation goes:

50 [lbs seedmeal] x 0.45 [percent protein] x 0.16 = 3.6.

Nitrogen can rapidly turn light-green leaves to a darker color, indicating more chlorophyll is present; more chlorophyll allows the plant to manufacture more sugar. So nitrogen is the most noticeable plant growth accelerator. Every plant protein molecule has a nitrogen atom in every amino acid in it. The most important plant protein is chlorophyll, the green pigment that converts sunshine into sugar; could anything be more crucial to plant growth than that? Dark-green leaves grown on fully mineralized soil can contain over 20% protein; about as high a protein content as in beefsteak, making chlorophyll more or less the ideal mainstay of human and animal health. Best of all, chlorophyll eaten raw is far more digestible than cooked animal flesh or legume seeds. If only there were some way we humans could enjoy life while eating nothing but raw leaves grown on fully balanced soil...we'd all live to be 150-year-old gorillas.

The highest yielding, high-plant-density, irrigated field crops can use about 200 pounds of soil nitrogen over their growing cycle. If that crop does not access 200 pounds of nitrogen (plus a sufficiency of everything else required to balance that N), it won't make a record yield. However, a bounteous crop of lettuce, carrots or beet roots needs only 80 pounds. Fruiting crops, like the Solanums and Cucurbits, use heaps of nitrogen while they are mostly making new leaves, but when they begin ripening a fruit load (or filling out potato tubers), vegetative growth slows, and little or no new chlorophyll is formed; so their need for nitrogen goes way down. But if you want to see giant cauliflower and broccoli, or celery with stalks up to your waist, you need to abundantly supply them with nitrogen (in balance with everything else, which basically means heaps of everything).

Target Level for *added* nitrogen: 100 lb/acre in a mixed vegetable garden that has not yet been fully balanced.

100 lb/acre = 0.25 lbs N per 100 sq ft

0.25 lbs N per 100 sq ft = 1½ quarts of feathermeal, or 3 quarts of seedmeal (the same quantity as in my COF recipe given in Chapter 4).

Large-scale farmers should plow in legume green crops for nitrogen and grow themselves some humus at the same time, but economics don't support this sustainable practice. Gardeners can do better. We can *aspire* to *gradually* eliminate all imported nitrates (including organic sources) as our soil achieves the kind of fertility that produces its own nitrates in abundance. Note that I stress *aspire* and *gradually*.

There are several biological nitrate sources we can encourage. The main one is nitrate fixation in legume root nodules by rhizobia, a microorganism inhabiting the nodules it provokes on the roots of legumes. *Frankia* (a type of bacteria, previously named *Azotobacter*) do something similar (and less intensely) for non-legume species. We're talking about the possibility of manufacturing quite a bit of nitrate. A well-grown stand of small-seeded broad beans can fix about 100 pounds of actual N/acre — while producing a good bit of new organic matter as well. Wild lupins, white or blue, make similar quantities of rhizobial nitrogen. Other legumes, like soybeans, lentils and chick peas, make less (60–80 pounds). Garden peas and ordinary beans do fix some nitrate, but not enough for even their own requirement; to grow well they need extra N from the soil, like any low-demand crop. If a legume crop forms seed, virtually all the nitrogen created through the entire growing season will be stored in that seed; there's very little left in the root system.

The presence of 100 pounds per acre of imported nitrate stops microbial fixation. Thus, nitrate fertilizer is addictive; feed the soil a lot of it, and your garden becomes dependent on nitrate fertilizer, be it chemical or organic, ammonium sulfate or chicken manure. Ideally, in veggie gardening, you'd use none of either kind — or just the smallest effective amount — and only on high-demand crops. Ideally, not practically.

A frequently repeated error in veggie gardening books says legumes release effective quantities of nitrates into the soil *while they are growing*. The truth is that virtually all rhizobial nitrates are immediately moved into newly forming leaves to make chlorophyll. If the whole plant later decomposes into the soil, it releases most of these nitrates for the following crop. The problem with all of this is that growing garden nitrogen with legume green manures requires allocating space to inedible crops that must grow for months and then be turned under. And then you have to wait two weeks to a month for them to decompose before starting a food crop on that ground. However, legume vegetation can be ripped out and put into the compost, allowing you to set seedlings directly into the root stubble without digging. Still, it grew there for months and was not edible.

Legume green manure crops rarely accord with most gardener's desires. Still, it's wise to plan on long-term rotations and legume green manure crops — if you have surplus space or can overwinter them after a summer crop. I do this. All my summer vegetables, the ones that finish when it already has become too late to start another food crop, are followed by small-seeded broad beans or blue lupins over winter, and then put to spring-sown crops a few weeks after digging the green crop in.

Frankia are of nearly equal value to rhizobia. Some types associate with the root zones of compatible crop species; some are free-living. The amount of nitrates *Frankia* fix varies, mostly according to soil quality. Farm consultants usually do not give the microbial creation of soil nitrates much importance because the quantity of nitrates created depends on there being plenty of high-quality soil humus, and in farm soils, there usually ain't. Soil-dwelling nitrate-fixing bacteria eat (decompose) organic matter. And they require abundant soil oxygen, which means that for them to work effectively, the calcium-to-magnesium saturation ratio must be in the right ballpark. They require a balanced abundance of all the usual plant nutrients.

We can aspire to creating garden soil that is entirely independent of concentrated nitrate fertilizer of any sort and yet still offer plants quite high levels. But it may take several years to bring soil to this degree of health and for you to educate yourself to that degree of understanding. I do not recommend withdrawing nitrate fertilizers all at once. Instead, encourage the garden to do its own nitrate production — initially, by importing 100 pounds per year, best in the form of a potent organic concentrate like seedmeal, feathermeal or fishmeal. Then each subsequent year, as your soil organic matter level comes up, and your mineral balance gets closer to being on target, give it less N. I suggest reducing nitrogen by about one quarter each year compared to the previous year: first year, 100 pounds; second year, 75 pounds; third year, about 60 pounds; fourth year, 45 pounds of N fed only to the most demanding crops, etc. Of course, this conversion must be accompanied by adding the highest possible quality compost for several years ongoing. If you have not yet learned to make great compost, don't stop importing nitrogen. If this year's reduced application doesn't produce the same rapid growth you enjoyed last year, then side-dress more.

In the same way that soil can produce it own nitrates, it is a real possibility that a perfectly orchestrated garden can produce enough organic matter to sustain its current organic matter level. Achieving that outcome would be a work of art I'd greatly admire. The moral of this tale: It is wise practice to limit nitrogen fertilizer to the minimum absolutely needed, so you encourage the garden itself to start becoming independent of the fertilizer sack. But on the other hand, don't shoot yourself in the foot trying to get there too quickly.

I divide the nitrate needs of garden vegetables into low-, medium- and high-demand (see sidebar for a list). Low-demand veggies need no more than 50 pounds of nitrogen per acre per crop — a small quantity many garden soils release by themselves. Before sowing low-demand crops, I suggest spreading one quart of seedmeal per 100 square feet, or half that, if you're using feathermeal. If that bed had already been given its yearly dose and has already grown one crop that was given plenty of N, then there's probably enough remaining in the soil for a following, low-demand crop. If I'm growing a medium-demand crop, I'll feed the bed two quarts of seedmeal, or one of feathermeal or fishmeal prior to sowing; if it is a high-demand crop, I give it three quarts of seedmeal, or half as much feather or fishmeal. No matter, be they low-, medium- or high-demand, all crops need the full-strength complement of all the other plant nutrient minerals.

- Low-Demand Crops: Parsnip, beet root, carrot, rutabaga (swede), kale, collards, beans, peas, turnips, winter storage radishes, herbs, horseradish, fruit trees and other small fruit, Swiss chard (silver beet), cereal grains of all types (except corn) and Jerusalem artichokes.
- Medium-Demand Crops: Tomato, pepper, eggplant, potato (sweet and Irish), Brussels sprouts, cabbage, kohlrabi, endive, lettuce, parsley, small salad radishes, pumpkins, summer squash, cucumbers, melons, mustard greens in autumn and most other Asian greens, okra, asparagus, field corn.
- High-Demand Crops: Broccoli, cauliflower, celery, rhubarb, winter squash (high-demand only if you have room for them to really run), mustard greens in spring, hybrid sweet corn.

Nitrogen in excess can interfere with phosphorus uptake. As N goes up, P must go up in accord. And high levels of P can interfere with zinc uptake. It's a complex puzzle, best not to bring it to the fore; best to keep your soil nitrate levels as low as is consistent with reasonable growth.

Although nitrogen is not on the worksheet, Matthew's target level for mixed vegetables should be 200 lb./acre N. Assume his existing soil organic matter releases 75 lb/acre. Therefore the deficit is 125 lb/acre N.

The Minor Nutrients

The textbooks call them "trace elements," but I resist diminishing zinc, copper, manganese and iron like that. A "trace" should be a streak that whizzes by almost before being noticed — something only vaguely present in minute quantities. Tiny. In this category, I put essential nutrients like vanadium and cobalt. A few grams in an entire acre is a gracious plenty, but complete absence of these elements means catastrophe for either the plants or for the animals eating them. Add a few ounces more than necessary of some of these, and the entire ecosystem is poisoned. And as long as we're on the subject, some trace elements are picked up and incorporated into plants, but may not be at all required for their successful growth. These include iodine and selenium. These two, however, are essential for human health and have to be in soil only at a few grams per acre in order to be present in your food. Most soils — the great majority of soils — are adequately supplied with micronutrients. But some regions are seriously deficient. Australia once converted huge expanses of useless land into productive farms simply by aerial broadcasting a few ounces of molybdenum per acre. If this is the situation in your region, your local garden center or extension office will know all about it. Routine use of kelp meal or a trace mineral fertilizer like Azomite insure against micronutrient deficiencies.

Zinc, copper, manganese and iron should be present at a level beyond a trace. The soil should hold zinc, manganese or copper in quantities of tens of pounds per acre, or, in the case of iron, a few hundreds of pounds in each acre, not just a mere trace. But major, minor, trace or micro, all nutrients are essential, maybe almost equally essential.

Iron/Manganese Balance

The target levels for these two elements does not rise and fall in strict mathematical accord with the TCEC. Still, light soil needs less than

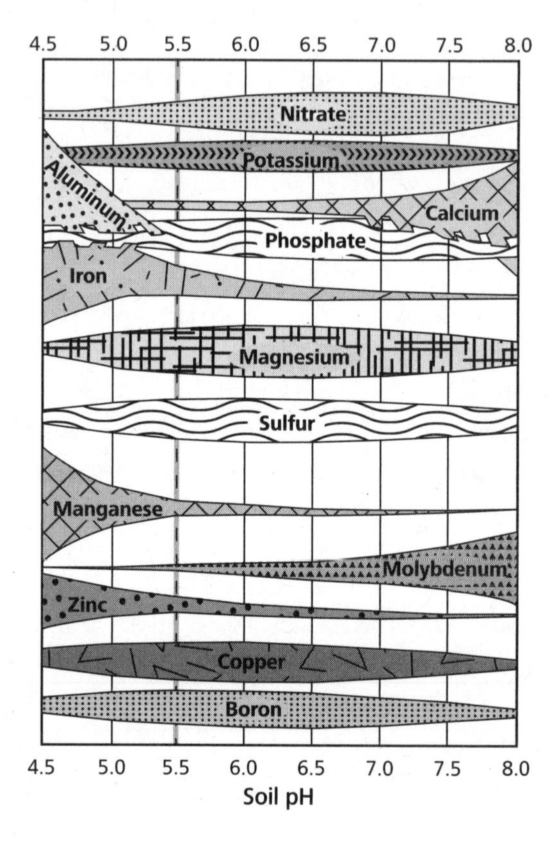

Fig. 5.5: *Availability.*

heavy soil does. Manganese/iron levels are interdependent; there should be at least one-third more iron present than manganese. Manganese/iron targets are uncertain. Every farm advisor has an opinion. Erica and I did much deliberating over the evidence before settling on the safe-yet-abundant levels we suggest in this book.

Iron (Fe)

Iron-deficient soils are extremely rare. It is normal to see M3 soil audits showing iron at 400 lb/acre or more. Sometimes plenty of iron may be present, but the plants exhibit deficiency signs because in some soils iron locks up into highly insoluble combinations. If you add iron without changing how that soil deals with iron (if that were even possible), it soon becomes unavailable. The usual reason for iron lock-up is a pH above 7.5. The best way to deal with plants that exhibit iron shortages is to lower the soil pH if possible, and to up the soil's organic matter level, which usually is more than possible in a veggie garden. Lowering the pH chemically unlocks iron. More organic matter provides more active sites for iron to attach itself to in an available form, and it increases the activity of the soil ecology, so it releases more iron and provides microzones of higher acidity around decomposing bits of organic matter. In these zones, there will be available iron. Meanwhile, an interim dose of iron sulfate will help produce healthy crops.

Minimum Levels for Iron

TCEC below 10: 100 pounds per acre.

TCEC above 10: 150 pounds per acre.

If your soil pH is below 7.6, and your soil test shows an iron level below the minimum target level, then use the pinkish form of iron — iron sulfate (ferrous sulfate) — to raise the level to that minimum. If soil pH is above 7.0, and you hope to lower it, do not add what may become unneeded iron to the soil. Instead, consider foliar feeding iron to your crops until such time as the soil conditions improve, allowing the iron, currently hidden, to reveal itself.

Manganese (Mn)

As with iron, manganese is rarely short, although Bill McKibben says loose, well-aerated, high pH soils are almost sure to need it. Of more concern is the risk of manganese toxicity in light, naturally acidic soil. Manganese differs from other nutrients in how strongly it reacts to changes in soil pH. While copper and zinc do increase in availability as pH drops, manganese becomes about 100 times more available as pH moves from 6.5 to 5.5.

When a very light farm soil (not garden soil) is fertilized, its exchange points rapidly get saturated with cations; accordingly, its pH rises to a comfortable range for plant growth. But the size of the pantry is so small that before the crop has grown completely, it may have so drained the nutrients held on the TCEC that the soil pH goes highly acidic. This prompts available manganese to reach toxic levels. Such soil lacks buffering capacity; ultra-light soil needs an organic matter ballast, like a fluorescent lamp needs one. A dose of cations (fertilizer) will rapidly re-elevate the pH and end the damage — if the crop is salvageable. So, I recommend building lower manganese levels for light soils than for heavy ones.

Unlike farmers, gardeners and homesteaders can elevate total cation exchange capacity, thereby providing a more effective buffer to stabilize pH and provide more constant nutrition. For this reason, I am not going to waste your time with a full recitation of symptoms of manganese toxicity or how various temperature and moisture conditions act and interact to up and lower manganese levels. Best we simply do not allow this problem to exist. Far easier to spread compost.

Manganese

TCEC below 10: Target Level is 55 lb/acre.

TCEC above 10: Target Level is 100 lb/acre.

As pH increases above 7.0, manganese goes relatively unavailable. Growers with high-pH soil, and especially with high-pH, light soil, can respond by banding manganese sulfate ($MnSO_4$) immediately below and/or beside plant rows at the rate of 10 to 20 lb/ac (10–20 grams/100 sq ft) $MnSO_4$. Or, a foliar feed of manganese can be used.

All crops require some manganese. Soybeans, lettuce, spinach, onion, potato, peas, beans, radishes and beets have a high requirement. If you added manganese this year, or if tested manganese levels barely reach my recommended minimum levels, you might try one foliar feed of manganese, Do it about the time the crop establishes a leaf canopy. If you get a growth response in a few days, then you needed to do it. If you get no response, there is no need to try it again. This suggestion holds true of any of the four "trace" elements.

Copper (Cu)

Copper and zinc probably do not have a relationship as nutrients, but it is convenient to reckon copper at half of zinc. At high concentrations,

Copper

Target Level is ½ zinc Target Level.

Maximum safe addition on light soils is seven pounds elemental copper per acre per application. Very heavy soils (TCEC over 20) with abundant organic matter may tolerate twice that quantity at one go.

copper sulfate is poisonous. Excess copper will suppress and/or kill soil microorganisms. But plants and the soil ecology both need *some* copper; it seems to have something to do with the immune function. Its presence certainly makes food taste better. The worksheet limits additions to no more than seven pounds actual copper per acre to avoid temporarily poisoning the soil ecology during the short time it take copper cations to attach themselves to the TCEC. Seven pounds of copper should provide a sufficiency even in soils that have next to none to start with. Copper sulfate may have to be added several years running before the background levels build up.

As soil pH goes up, copper (and zinc) become less available. On calcareous soils, it may be necessary to foliar feed them. Copper sulfate in solution can be harsh on leaves, so when sprayed as a foliar, it must be diluted to half the concentration of the other sulfates.

The soil's ability to adsorb copper has a lot to do with its organic matter content. If your heavy-soil garden has high organic matter and is seriously copper (and/or zinc) deficient, you can double the application limits on copper and zinc.

Zinc (Zn)

Zinc

Target Level is 1/10 the target for phosphorus.

There is an application limit of 14 pounds of elemental zinc per acre per application. However, very heavy soils (TCEC over 20) with abundant organic matter may tolerate twice that quantity at one go.

North Carolina State University says that zinc is the most commonly deficient plant nutrient, and not just in North Carolina. Shortages most often appear in leached, acidic, sandy soils, of which North Carolina has an abundance. Zinc uptake can be suppressed by high levels of available phosphorus (and vice-versa), which is part of why this book links zinc and phosphorus targets and limits application amounts for both elements.

Excesses

Any time the actual level of an element exceeds your target level, you have an excess. Matthew's analysis (shown in Figure 5.6) shows excess phosphorus and zinc. If the excess is slight — no more than 10% over the desired level — don't bother about it. Let it be. Ten percent one way or the other is within the level of accuracy we're working with. If the excess involves anything other than the four major cations, there is nothing I know of you can specifically do about it except to not add more. If a huge excess in one item is causing a deficiency in another, sometimes you can intentionally create a balancing excess. But to deal with this, you'll probably need advice from an experienced analyst. If it is an excess cation, gypsum will reduce it. If it is an excess anion, the passage of time will lower it. But anions pose no immediate threat; crops usually do not noticeably react to surplus sulfur or phosphorus.

Excesses in any of the four major cations can be adjusted by taking advantage of how cations behave naturally. Cations can be seen as being in an unequal competition to hook up with the fixed, permanent number of the soil's negative charges or exchange points. Whichever cations are the most concentrated in the soil solution tend to displace those on the exchange points. Thus, if we intentionally put in a big dose of one cation, it will reduce the levels of some others. In the glossary given in Chapter 1, I set you up not to be shocked when I tell you again that some cations are *divalent,* meaning they carry two positive charges, while others are *monovalent,* carrying only one positive charge. A divalent cation connects to two exchange points and thus, holds on with greater energy compared to a cation with only one positive charge. So a soil solution holding a high concentration of calcium or magnesium, which are divalent cations, will readily replace exchangeable

Acid Soil Worksheet
U.S. Measurements

Name *Matthew*
Plot or Field *Unfarmed*
Date of Test *2/3/2012*

Sample Depth 6 inches All numbers on this worksheet assume a six inch sample depth

TCEC *13* **TEC:** If CEC is below 10 it is a "light soil." Over 10 is "heavy soil."

pH *5.8* **pH:** If pH is 7.0 to 7.6, go to the Excess Cations Worksheet. If pH exceeds 7.6, you may have calcareous soil.

Organic Matter % *5.4* **OM:** Target over 7% in cool climates. South of the Mason-Dixon Line target over 4%. Assume an approximate release of 15–25 lb nitrogen per 1% OM. Varies with temperature, moisture and soil air supply. N = 0.22 x NO_3

Element	Actual Level	Calculating Target Level Pounds per acre	Target Pounds per acre	Deficit Pounds per acre
Sulfur **S**	ppm *26* lb/ac *52*	**S = ½ Mg (Target Level)** until there are no cation excesses; then **you *may* Target S=1/3 K**	*187*	*135*
Phosphorus **P**	P_2O_5 *954* P = *420*	**P = K (Target Level)** Calculate using actual P, not phosphate. P = 0.44 x P_2O_5	*350*	*Excess*
Calcium **Ca**	lb/ac *3292*	**TCEC x 400 x 0.68 = Target Level**	Minimum target 1,900 lb/ac *3536*	*244*
Magnesium **Mg**	lb/ac *210*	**TCEC x 240 x 0.12 = Target Level**	*374*	*164*
Potassium **K**	lb/ac *125*	**K is proportional to TCEC: see chart**	*350*	*225*
Sodium **Na**	lb/ac *73*	**TCEC x 460 x 0.02 = Target Level** Be certain of good water quality before adding sodium	Do not exceed 160 pounds *120*	*47*
Boron **B**	ppm *0.63* lb/ac *1.3*	**B = 2 lb/ac if CEC below 10 = 4 lb/ac if CEC above 10**	Do not exceed 4 pounds *4*	*2.7*
Iron **Fe**	ppm *206* lb/ac *412*	**Fe = 100 lb/ac if CEC below 10 = 150 lb/ac if CEC above 10**	*150*	*Excess*
Manganese **Mn**	ppm *8* lb/ac *16*	**Mn = 55 lb/ac if CEC below 10 = 100 lb/ac if CEC above 10**	*100*	*84*
Copper **Cu**	ppm *5.16* lb/ac *10.3*	**Cu = ½ Zn (Target Level)**	*18*	*8*
Zinc **Zn**	ppm *31* lb/ac *62*	**Zn = 1/10 P (Target Level)**	*35*	*Excess*

Potassium Target Levels

TCEC	Pounds	TCEC	Pounds	TCEC	Pounds
		16	390	28	493
		17	400	29	500
		18	410	30	507
7	255	19	420	31	511
8	270	20	435	32	515
9	290	21	443	33	519
10	310	22	451	34	523
11	320	23	459	35	527
12	335	24	463	36	531
13	350	25	475	37	535
14	365	26	481	38	539
15	380	27	487	39	543

If TCEC is lower than 7, use value for 7. If it is over 39, use value for 39.

	One acre, six inches deep weighs	One hectare, 80 mm deep weighs
1 meq Calcium	**400 lb**	**400 kg**
1 meq Magnesium	**240 lb**	**240 kg**
1 meq Potassium	**780 lb**	**780 kg**
1 meq Sodium	**460 lb**	**460 kg**

1 ppm = 1mg/kg = 2 pounds/acre = 2.24 kg/hectare

Date of Issue: 07/07/2012

Fig. 5.6: *Page 1 of Matthew Preston's Acid Soil Worksheet showing calculations for target levels and deficits, completely filled in.*

(cations already on exchange points) potassium or sodium, which are monovalent cations, with calcium or magnesium. One last factor comes into play: some cations naturally cling with more energy than others. This has nothing to do with the number of attachment points. Soil scientists believe they understand the physics behind this, but to transform soil, you do not need to know the why of it. Only the how and the what. The bottom line is that calcium clings harder than magnesium can, and potassium holds to clay more tightly than sodium can. You see this demonstrated in Albrecht's saturation percentage targets: mostly calcium, one-seventh the quantity of magnesium as calcium, one-quarter the potassium as magnesium; one-half the sodium as potassium. A high concentration of calcium cations in the soil solution will knock magnesium off the TCEC; a high concentration of magnesium in solution will displace potassium; similarly, potassium displaces sodium.

An abundance of calcium cations in the soil solution creates a cation cascade, lowering the levels of the other three majors, moving these cations back into the soil solution. But excess cations can only be eliminated if the soil drains freely and gets leached. If it rarely rains enough to rinse out your soil solution, you still can heavily irrigate free-draining soil and thereby leach excess cations.

The cascade of cation replacement is usually written shorthand this way:

Ca > Mg > K > Na

Calcium replaces magnesium, which replaces potassium, which replaces sodium.

Excess Calcium

The most frequent home-garden excess (if any calcium level can actually be a damaging excess, and about this I am still uncertain) comes from liming. A slight lime excess, a few tons per acre, is usually not hard to repair. If the excess is from dolomite, though, you may also have a magnesium excess to handle. But do not stress: both of these excesses will work themselves out at the same time.

The way to reduce calcium is with *scant* applications of agricultural sulfur, never more than 100 pounds elemental sulfur per acre per year. Sulfate anions aggressively combine with any available cation, forming soluble sulfate salts. Since the most frequently found cation in a soil with excess calcium will be calcium, most of that sulfur will form calcium sulfate (gypsum), which is soluble and leachable. High levels of soil-manufactured calcium sulfate will gradually reduce the saturation levels of the other cations, but the main effect will be on the largest concentration of cations — calcium. Ag sulfur gradually and gently reduces (mainly calcium) excess as long as you allow the sulfur levels to be as high as the sulfur target suggests.

This works out to have a double benefit. The leached calcium and other cation nutrients settle in the subsoil, increasing its pH, and saturating it with plant nutrients. The effect of this is similar to putting a fast-growing seedling into a larger flower pot.

In the event you run out of excesses to leach, you may elect to continue adding the soil's remaining sulfur requirement in the form of gypsum, which will then continue to restock the subsoil with cations. It'll take some years, but the result can be amazing. I remember well the great pleasure Annie and I experienced when the extremely flavorsome beet variety we had been growing for a few years suddenly doubled in flavor and sweetness. This happened in my pre-remineralization days, when I used COF. It had taken three years for enough calcium and other plant nutrients to recharge my subsoil to the extent that the beets, primarily subsoil feeders, began to get sufficient nutrition.

Handling large excesses of calcium, stemming from either more than one or two tons per acre of excess lime or from natural causes, are another matter and are discussed in the next chapter.

Excess Magnesium, Potassium or Sodium

It is highly unlikely that any acidic soil has excess sodium because sodium has a powerful effect on soil pH; if sodium were much in excess, the soil would not be acidic. Erica Reinheimer's garden began with a slightly acidic soil having magnesium saturation way over 12% and a large calcium deficit; this is not unusual. The excess magnesium pushed her pH far higher than it otherwise would have been. This sort of excess is resolved by adding enough ag lime to satisfy the calcium deficit and then adding gypsum up to the soil's sulfur target, even though that gypsum apparently pushes calcium over your target. Gypsum will not raise soil pH. And it doesn't always increase calcium saturation; ag lime does that with greater certainty. But it *will* bump excess magnesium and potassium off the clay. And it will increase available calcium.

Chapter 7 addresses handling neutral pH and calcareous soils that, almost by definition, have large excesses.

Moving Along

If you are like the great majority of readers, you have acidic soil, and you're now ready to learn how to work out a list of materials for your own soil prescription. If there is excess magnesium (or potassium or sodium) in your soil, you should have confidence that the matter will get sorted out over the next few years. Once you've done your own calculations for Matthew's example soil test, you will be ready to work out your own soil analysis.

Chapter 6

Le Batterie de Cuisine

Calcium (Ca)

Limestone is mostly calcium carbonate, a mineral that is properly termed "calcite." Before we had the steel to grind it with, calcite was cooked in a lime kiln using firewood. In the kiln, it crumbled to calcium oxide (quick lime); then water was added, turning it into slaked or hydrated lime (calcium hydroxide), which naturally combines with carbon dioxide to form calcite again, but this time, in fine powder form. Limestone deposits vary in purity. It is not uncommon to find 99% pure calcium limestone. Most limestone contains some dolomite (calcium-magnesium carbonate), but usually not much. When it contains quite a bit of dolomite, we call it dolomitic lime; sometimes it is pure dolomite. Legally, to be labeled "limestone," the rock needs be no more than 90% pure calcite. When the magnesium level is only a percent or two, it can be ignored. (An interesting aside: We have a small deposit of lime on Tasmania "contaminated" with 5% phosphorus. And sometimes lime comes with useful amounts of boron or other elements.)

Tasmania doesn't have many rules, and our impoverished government can't afford to enforce the rules it has. So when I buy garden lime, it usually comes in an unmarked plastic sack with a homemade wire twist-tie holding it shut. The grind size is...whatever it is. There is no analysis. But in North America, you should see an analysis on the bag showing the purity (pure would be 39%–40% calcium) and magnesium content (if any). If you were a farmer buying lime by the ton direct from the nearest quarry, you'd be supplied with an analysis if you asked for one.

Ag lime is graded according to particle size, which amounts to how quickly it will dissolve in soil. So, specifying the grind you want can be quite important. Lime is graded according to the percentage of it that will pass through a sieve of a fixed size. If the lot is labeled #10, it means that all the material will pass through gaps slightly smaller than $^1/_{10}$ inch. Another common grind is #65, meaning the material passes through a screen with 65 lines per inch — so only particles smaller than $^1/_{64}$ inch get through. It is not uncommon to find #100 ag lime, which resembles the grit on fine sandpaper; there are superfine grinds on offer as well.

The finer the grind, the more the cost. Keep in mind that when all the material passes a #10 screen, a sizeable fraction of it will pass a #100. Farmers spread coarsely ground lime thickly, expecting that the finer particles will dissolve quickly and bring things to balance while the larger-sized chunks will go on replacing lost calcium for a decade or more.

Neil Kinsey says that any fragment of limestone that will pass a #65 screen will dissolve in the soil within three years. When balancing soil, we want calcium to appear quickly, so I urge you to use #100 agricultural lime, or #65, if you can't find #100. Even the finest lime will mix uniformly and easily into seedmeal, making it easy to spread.

Oyster shell lime is finely ground, high-purity calcium carbonate that carries a few percent of a broad range of other micronutrients because it comes fresh from the ocean. My best guess (formed without personal experience with oyster shell lime) is that oyster shells used as fertilizer are not worth any more than ordinary high-calcium #100 ag lime, but the oyster shells come with a higher price tag.

You can also buy marble dust, an ultra-fine-grind limestone. Peaceful Valley Farm Supply sells one so fine that it can be mixed in water and sprayed through a nozzle. It has a quick reaction in soil. This stuff mixed in irrigation water could rescue calcium-starved containerized plants. But it costs several times the price of ag lime, which I find a bit off-putting — they're probably just bagging up waste dust from grinding and polishing slabs of marble and selling it at a very high price. I think ordinary #100 ag lime is ideal for gardens; #65 will do.

When exposed to air, finely ground ag lime has a tendency to form large, hard chunks that cannot be uniformly spread unless they are broken up thoroughly, which isn't easy to do. So it is sensible practice to store opened sacks of #100 lime in air-tight containers. It may also be a good idea not to stock up on this material.

Be aware that OMRI-approved phosphate fertilizers contain considerable amounts of calcium. These "freebie" sources need not be accounted for when working out a list of amendments.

The last, but absolutely not the least form of lime, is gypsum, which is calcium sulfate. Gypsum can be a byproduct of chemical manufacturing, but it is also a naturally occurring mined rock (of concern, if you seek formal organic certification). Some types are more soluble (quicker releasing) than others. In the medium-term, the difference between medium and high solubility makes no difference. Gypsum is soft; it dissolves into moist soil rather quickly. Gypsum can be considered primarily as a source of readily available calcium or primarily as a source of sulfur, depending on the circumstance. Gypsum has chemical variants, so you will see different analyses. Best to reckon it is 23% calcium and 18% sulfur.

Magnesium (Mg)

The most economical source of magnesium is dolomite lime — if you also need calcium. A typical dolomite analysis is 27% calcium and 11.5% magnesium. But there is a downside to dolomite: it is much harder than calcite, so it is slower to dissolve in soil. Because it is so hard, you may not see the full result from #65 dolomite for years. This means, if you forget about having added dolomite the previous year or two and respond to a soil audit calling for more magnesium with more dolomite, you may overshoot a few years later.

Pro-Pell-It is a brand of granulated (prilled or pelletized) ultra-fine-grind, high-purity dolomite lime that can be counted on to break down as rapidly as dolomite can. Nearly 100% of Pro-Pell-It passes a #100 screen. I recommend it.

Because dolomite dissolves slowly, you might continue to experience magnesium deficiency symptoms for a year or more after adding it. But be aware that a soil audit showing a magnesium deficiency may not mean the plants growing there will be deficient. Plants have a surprising ability to discover magnesium even when a soil test fails to. Gross magnesium deficiency is recognizable by whitish or very light-green streaks on the leaves; they appear because the leaves have failed to form sufficient chlorophyll. In the event you suspect a deficiency of magnesium, an easy way to confirm it is to foliar feed one heaping tablespoon of Epsom salts in one gallon of water. If the next sets of leaves are properly green, magnesium was deficient. In that case, continue to foliar feed Epsom salts, probably once every few weeks. In my opinion, Epsom salts are too costly to routinely mix into soil.

Two similar, naturally occurring mined minerals, sold as K-Mag and langbeinite, dissolve quickly in moist humusy soil. Each contains magnesium, sulfur and potassium, in similar proportions. However, as sensible as these products are, they're not often used; on most occasions when magnesium is called for, potassium is not needed. K-Mag contains 11% Mg; langbeinite provides 12%.

Michael Astera considers the best and often the cheapest source of magnesium to be magnesium oxide (MgO). For inexplicable reasons of the sort that career bureaucrats grasp easily, but I lack the mentality to comprehend, MgO is not approved for organics. However, MgO is sold as an animal feed supplement and also in health food stores. MgO is 50%–55% elemental Mg. It needs to be very finely ground to have a rapid effect.

Potassium (K)

The easiest to find, cheapest, approved-for-organics K fertilizer is potassium sulfate (K_2SO_4). The bag label says it contains 50% K_2O, but that is only 41% actual K. It also contains 17.5% sulfur.

K_2SO_4 was once made by reacting natural, mined potassium chloride, KCl, with sulfuric acid. J.I. Rodale considered KCl suitable for organics because it is a natural, mined substance; he considered K_2SO_4 unacceptable because it is artificial. These days, the most common production method uses natural rocks, like langbeinite, ground finely and then carefully rinsed with salt solutions to produce pure K_2SO_4. A similar process is used to make it at the Great Salt Lake in Utah. If you take some time to shop around, you can find OMRI-approved, naturally mined potassium sulfate that is useful, but it comes with a minor liability: not being pure enough to be instantly water soluble, it cannot be used for foliar spraying or to make liquid fertilizers.

Greensand (also known as *glauconite*) is a natural, mined, very slowly releasing potassium source (it also contains a range of trace elements). I have found greensand analyses ranging from 3% to 7% potassium. This is a surprisingly wide range to find in an analysis. Perhaps some garden writers confuse K with K_2O. Or it may be that greensand deposits vary in K concentration. Greensand is a sedimentary, clayish rock with a built-in cation exchange capacity that slightly increases the TCEC of light soils. It releases potassium very gradually, making it useful for building a soil's potassium reserves, but not as a quick remedy for a potassium deficiency. If you were to use enough greensand to handle an immediate need for a good bit of K, you would need a huge amount — the cost would be prohibitive. If you were to routinely spread a ton/acre every year into the homestead garden, your grandchildren might not ever need to use K fertilizer. If you're contemplating investing in future generations, remember that greensand contains 3% Mg. The magnesium shouldn't release any faster than the potassium, but do you really want that much Mg sitting around in your soil's reserves?

Sometimes granite dust is used as a potassium fertilizer. It could be a valuable farm soil amendment if it were available at prices similar to ag lime straight from the quarry. Be aware, though, that there are two basic kinds of granites: one mostly has orthoclase feldspar in it; the other mostly plagioclase feldspar. Orthoclase has the potassium content; plagioclase has mostly calcium in it. Orthoclase is pinkish in color; plagioclase is white. Get an analysis!

The minerals in langbeinite rapidly become available in soil. It contains about 19% K, 12% Mg and 23% S. K-Mag is a similar product, with 18% K, 11% Mg and 22% S. However, except for a new garden in highly acidic ground, it is a rare soil that calls for magnesium and potassium at the same time.

If a soil also needs calcium, wood ashes can serve as potassium fertilizer. An average analysis might be 1-2% P, 3-8% K and 20-40% Ca. However, wood ash mineralization varies greatly. Ashes also provide small quantities of whatever minor nutrients and trace elements the tree picked up, typically 1% Mg and useful quantities of iron and manganese. There may be traces of other plant nutrients, like copper and zinc, but usually the concentration is too small to be meaningful. Ashes are tiny particles that react rapidly in soil. They make an excellent compost heap ingredient. But be careful. Never spread more than 1 pound per 100 square feet on soil or use more in your compost pile than 10 pounds of ash per 1,000-square feet of intended finished compost coverage. At high levels, ash can be toxic, even to a compost heap. In case you haven't got the message yet, I suggest you do not use wood ashes unless your soil tests as needing calcium and magnesium and potassium. Or if you're desperate. In my opinion, the best place for a wildcard like wood ashes is back into the forest from whence they originated (and to save trees from shock, spread it thinly).

Potassium chloride (KCl) is the cheapest form of K fertilizer. However, KCl puts far too much chlorine into the soil and should be avoided, not so much because chlorine is toxic — in small quantities, chlorine is an essential plant nutrient — but because of what chlorine does to soil fertility. Like sulfate, chlorine combines with calcium, forming calcium chloride (CaCl). The chloride of calcium is far more soluble than the sulfate, and it leaches readily. You can assume that for every pound of potassium chloride put into the topsoil, a pound of calcium leaks out of the subsoil. There are farm soils, especially sandy ones, where potassium chloride was long and heavily used to grow vegetables; there is so little calcium left in those soils that crops do not grow at all. Yet there is so much potassium present that the soil pH is above 7.0. Tiedjens's book *More Food From Soil Science* has much to say about these circumstances and their remedy through heavy liming.

KCl is also the "K" in most garden center NPK chemical blends. So these, too, should be avoided.

Sodium (Na)

It is easy enough to buy a bag of mined rock salt; any type of sodium chloride will put sodium into your ground. However, buying the right sort of sea salt will also get you a broad supply of trace elements. Take warning: there is so-called sea salt, and there is real sea salt. I say so-called because most of the sea salt on sale has had its other (valuable) elements skimmed off the bottom. Sea water contains a enormous range of salts, sodium chloride being the most abundant by far. As sea water evaporates, the various salts in it crystallize out of solution in sequence. Sodium chloride is amongst the first to crystallize out. The heavier salts solidify at a higher concentration than sodium chloride does; so, after the sodium chloride has been removed, the heavy "liquor" that remains is sent off to a chemical plant for separation into more valuable compounds.

When making genuine sea salt, all the water is evaporated without first skimming off the table salt, so all the minerals originally present in the sea water are present in the salt. Usually this kind is not pure white, and it costs more. For salt as a seasoning, I suggest buying the genuine kind. For gardens, try Redmond Natural Mineral Salt. Sea salt is about 35% sodium by weight.

Phosphorus (P)

Four kinds of natural phosphorus fertilizer are allowed in organic farming. All of them carry a calcium component. (Note that the calcium in hard rock phosphate [HRP] is so insoluble as to be insignificant; but so, too, is most of the phosphorus in HRP — unless you compost it first.) There is one synthetic phosphate fertilizer I recommend for use on calcareous and sometimes on neutral soils because it creates small zones of useful acidity when it dissolves.

Bonemeal

Two kinds are on offer: fishbonemeal and ordinary steamed bonemeal, a slaughterhouse byproduct. Both sorts analyze more or less at 3-15-0 and have the advantage of being quick to release phosphate. They're also about 30% calcium and nearly 6% sodium, so if your soil has too much of either, bonemeal is not the best option.

Hard Rock Phosphate (HRP)

Hard rock phosphate usually holds about 30% phosphate (13% phosphorus), of which 3% is available; the remaining 27% is quite unavailable and should not be counted. So, be aware, if you're planning on using HRP: its available P is a mere 1.3%. It takes a strong soil ecology many years to release more than the original 1.3% in an acidic soil. And in neutral or alkaline soils, HRP brings little benefit. On the other hand, it is the least costly natural phosphate. If a soil's insoluble phosphorus reserves are brought high enough, sufficient phosphorus will be naturally released to grow good crops. But spreading HRP at 4–6 tons per acre (quite a few times) is rather more costly these days, now that we are post-peak phosphorus. I would not use HRP if I could get soft rock phosphate. When you read what I have to say about monoammonium phosphate, you might not ever want to use HRP; if you still do, though, you should be mighty fussy about reading the analysis that comes with it.

The best way, by far, to use rock phosphate — soft or hard — is to first mix it into a forming compost heap where it will be digested by energetic bacteria and fungi. Second best is to blend it into damp, finished or nearly finished compost. Allow it to marry into the compost for a month before spreading it. Biological phosphorus digestion during composting is similar to what should happen in soil, but it happens much more effectively in a heap. Fermenting HRP this way for just a month or two may more than triple the level of available P.

Soils carrying excess available phosphorus usually arrive in that condition after multiple, perhaps excessive spreadings of animal manure. That phosphorus was, like the P in a compost heap, already part of organic matter before it was put into the soil. But, when phosphate fertilizer is simply spread right out of the sack and tilled in, it has a strong tendency to vanish from the next soil test because it becomes chemically insoluble before the organic fraction can assimilate it.

If you're gardening neutral soil or, even more so, in calcareous soil, the best way to use phosphate-fortified compost is to band it, making a high concentration immediately below the plant. The compost itself creates an acidic zone of superfertility where the phosphate within it won't rapidly become unavailable. I fully explain banding in Chapters 8 and 9.

Guano

Guano is sometimes natural, fossilized sea bird manure, mined in Peru. It is also sourced from bat caves. Sometimes guano contains as much nitrate as phosphate. Sometimes it is what is termed "high phosphate" guano, which is the sort I have used. It contains only 1% or 2% nitrogen — or none at all. The phosphorus in guano is reputed to be far more available than that from any other natural source. I have seen shockingly expensive forms of "micronized" guano intended for foliar feeding or mixing into liquid fertilizers. Because of its high price, guano is ideal only for potting mixes or other horticultural applications.

Soft Rock Phosphate (SRP)

I strongly prefer SRP over HRP. It contains about 9% elemental phosphorus (20% phosphate), a good deal of silicon, and about 20% calcium. All of these are held in a colloidal clay suspension, allowing the phosphorus in SRP to become available far more readily in soil than it can from HRP. Best, SRP's phosphate content will not lock up with calcium or iron as readily as the phosphate in other phosphorus fertilizers. Although HRP does include a bit more actual P for your buck, most of that hard phosphorus will feed those who come after you. SRP will mostly feed you and yours. It is the only natural rock phosphate that is effective in high-pH soils.

Some soils don't hold much quartz; consequently, the plants growing in such soils may be short silicon (Si). Much unappreciated by agronomists because it usually is abundant, silicon is as essential to plant functions as boron is. Short of spreading finely powdered glass or fine quartz sand, SRP is the only significant source of available silicon I know of.

The very best way to use SRP is to first marry it with compost. One 50 pound bag of SRP generously provisions about 1,000 square feet of garden with about 200 pounds total P per acre. This quantity of SRP is about the right amount to blend into one cubic yard of finished compost, which is about the volume it takes to cover 1,000 square feet about a quarter-inch thick. It is also about the right amount to blend into 2 to 3 cubic yards of starting volume when building a new compost heap. When SRP is mixed into a heap at the beginning, the heap will heat up faster and finish quicker because the compost ecology needs P as much as your garden plants do.

I have attempted to find out the CEC of the clay in SRP. To no avail, so far. I suspect someone with a light soil would get better compost by adding SRP because of the clay in it.

MAP (Monoammonium phosphate)

This is one synthetic fertilizer I urge you not to shy away from if it suits your circumstances. I suggest MAP when there is a need for phosphate, but the soil has a pH over 7.0, MAP at 23% actual P, is the best way to build phosphate levels in calcareous soils. It is also the cheapest P fertilizer I'm willing to use. MAP does not shock the soil ecology like DAP (di-ammonium phosphate) does. DAP is a harsh substance that is difficult for soil microlife to handle. DAP also carries too much N with it compared to the amount of P it delivers. I suggest you don't consider using DAP for any reason.

Monoammonium phosphate gradually dissolves into soil. While releasing phosphoric acid, it creates a small zone of moderate acidity around each granule that, in calcareous or alkaline soils, can maintain enough available P to get a good growing result. Microzones of acidity also help trace elements become more available. MAP contains one ammonium cation, NH^{4+}, that provides the soil with about 12% actual N by weight. So, if you're adding enough MAP to provide 175 pounds of P/acre, you're also adding 90 pounds of N; that is all the imported nitrogen a food garden needs for the year.

If you made a truly objective comparison — uncolored by prejudice against synthetic fertilizers in general — weighing the ecological costs and benefits of using MAP compared to using the HRP it was made from, you might conclude that MAP is a greener product, or at least, in the same ballpark. Monoammonium phosphate is made by rendering hard rock phosphate into the form of phosphoric acid; this also purifies it. Phosphoric acid is made in one of two ways, for two distinct purposes: one process leads to food-grade acid; the other, to fertilizer grade. The food-grade product is used in things like fizzy drinks. The fertilizer-grade acid is reacted with synthetic ammonia (synthesized from natural gas), creating MAP.

First of all, if judged by the effect it creates on the crop being fertilized, then HRP (at 1.3% available P) is only one-twentieth as potent as MAP, so you have to haul and spread up to 20 times more HRP to get the same immediate result. Transport costs are a major consideration in a world of post-peak oil. So is the energy cost of spreading fertilizers. HRP also contains a good deal of fluorine; you might not want to have this highly toxic element at high concentrations in your soil. MAP has been purified, it contains no fluorine. On the other hand, Hugh Lovel, a bio-dynamic advisor specializing in dairy farms, asserts that fluorine is one of the few elements capable of solubilizing silicon, so it's possible having some fluorine in the soil is all to the good. Depending on which deposit it came from, HRP can also contain far too much cadmium for comfort; cadmium, even in tiny doses, is a truly poisonous element. I've seen uranium on HRP analyses, averaging about a half pound of actual uranium to a ton of rock phosphate. All this toxic (and radioactive) dross — fluorine and cadmium and maybe uranium — is left behind when making MAP.

If we want more phosphorus in our farm soils, we must use HRP in one form or another because planetary reserves of soft rock phosphate and guano are so limited.

I anticipate a transformation in what is being sold at garden centers and farm suppliers as more gardeners begin seeking the full range of OMRI-approved materials. If my book proves to be effective at elevating awareness, some gardeners will be looking for MAP as well. To find MAP now, you probably have to contact a farm supply or major fertilizer dealer.

One last thing: if you're an organicist who is growling at me right now, please have a read of Donald Hopkins's book *Chemicals, Humus and the Soil*. He presents strong arguments that just might change your thinking on the subject.

Sulfur (S)

Sulfate fertilizers provide sulfur as a side-benefit. The amounts brought in along with potassium, manganese, copper and zinc can be substantial, especially if K is being boosted in the form of K_2SO_4. Gypsum also contains sulfur. In the event the quantities of other fertilizers called for are not large enough to supply a soil's sulfur requirement, *and* you have excess calcium to reduce, you can use agricultural sulfur, a finely ground yellow powder with a slight sulfurous odor. If you choose not to lower calcium levels, use gypsum to fill any additional sulfur requirement. When buying elemental sulfur, its particle size is important. When finely ground and well-distributed in moist, warm soil, pure sulfur is biologically converted to sulfate within one or two months. Ground coarsely (or in lumps), it can take years to fully react. The release of sulfur is temperature dependent. At 70°F the soil reaction is slow; it goes most rapidly over 85°. It virtually stops at 50°.

The best brand of agricultural sulfur I know of is Tiger 90, a superfine, quite pure sulfur that has been combined with 10% bentonite clay and compressed into stable granules that spread easily and uniformly. The granules disintegrate when they react with soil moisture. I have had satisfactory results in my own garden using ordinary ag sulfur, but I'd use Tiger 90 if it were sold in Tasmania.

Boron (B)

The easiest way for home gardeners to obtain boron is to use laundry borax from the supermarket. It assays somewhere between 9% and 11% boron by weight. I suggest you reckon it at 10% purity. Borax is a natural, unprocessed substance currently mined in the Mojave Desert in California. Boron is so powerful, and so little of it is required, that it should be thoroughly blended with other fertilizers so it gets distributed uniformly. Borax also can be dissolved in water and sprayed on the soil (it needs to be taken up by the roots, not the leaves). Some crops experience boron toxicity when the soil holds excess boron, so take care when measuring and spreading boron. Do not add more than two pounds per acre actual boron per application; that much should be more than sufficient for the year. Other forms of boron fertilizer may be more concentrated than borax, so be aware of that if you use Solubor or other agricultural boron supplements.

When measuring borax, two pounds per acre actual boron comes to 20 pounds per acre borax (at 10% B); or 20 grams (4 teaspoons) per 100 square feet. Six teaspoons of any salt is about the maximum quantity to dissolve in one gallon; more than that and you risk damaging plant leaves. If boron does get sprayed on leaves, you need to wash it off promptly, but you don't have to panic. Foliar boron won't hurt the plant (at the concentration I prescribe), but it does no good when not taken in through the roots.

Nitrogen (N)

All OMRI-approved forms of nitrogen fertilizer but one require biological breakdown to release their nitrate content. (Bloodmeal is the only one that is water soluble). When estimating how much nitrate will be released from an organic fertilizer, you could look it up in a gardening book, but you'll find a lot of variability in what books report. With oilseedmeal, you'll find variation from lot to lot and from type to type; to know accurately about seedmeals, you have to first see the label on the bag. Divide the amount of protein shown by 6.2 or multiply it by 0.16 (you get the same result either way). The law requires animal feeds to be labeled with their protein content. A 50-pound sack of oilseedmeal labeled 45% protein will release 7.2% nitrogen (45% divided by 6.2 = 7.2); So, the total amount of N in the 50-pound sack is 3.6 pounds (50 pounds times 7.2% = 3.6 pounds).

The release of nitrate from all organic sources, be they concentrates, compost or manure (except bloodmeal), is temperature dependent. Typical of biological chemistry, the rate of release doubles with each 19°F (10°C) increase in temperature. So, if you're in a place where the snow flies in winter, be aware that the rate of nitrogen release doubles from early spring, when soil temperature may be 40°F, compared to the amount released a month or two later, when it has reached 60°. In much of North America, the soil peaks around 80°F. At that temperature, the nitrogen release rate doubles again, to four times what it was in early spring. Thus, to get spring crops growing fast, it usually takes a bit of applied N — no matter how fertile the soil is otherwise.

Manure

Organic-certification bureaucrats no longer allow tilling in raw manure unless a long time passes before growing a food crop. This may actually be sound practice, not just bureaucratic muscle-flexing; raw manures can give vegetables off-flavors, and there are health concerns as well. So, unless you are in desperate straits and unable to obtain anything else, please do not use crude manure as nitrogen fertilizer.

I suggest you do not count on most animal manures to supply adequate nitrogen for demanding vegetables (exceptions being composted chicken manure that comes with an analysis and manures produced on your own remineralized, balanced property). Manures from livestock can be no more potent as fertilizer than their food was nutrient-dense. Anyone willing to sell manure or give it away probably does not value it highly, so is not likely to have handled it in such a manner as to insure against rapid loss of nitrates. Gardeners are often provided with statistical tables that supposedly show the nutrient contents of assorted animal manures, but when fresh manures are heaped up without making proper compost of them — even just for a week — much of their original nitrates are given off as ammonia gas. A similar thing happens when fresh, moist manure is not promptly gathered and is, instead, allowed to dry out in the sun. I suggest that you entirely discount the supposed nitrate contents of any manure except bagged chicken manure compost that comes with an analysis saying nitrogen is in excess of 3.5%. Chicken manure compost is great stuff. It can really grow things. Another source of manure you can count on to make excellent compost is rabbit manure from your own hutches — but only if there is enough straw (not sawdust) under the rabbits to soak up their urine.

When Albert Howard was researching how to make powerful compost, he knew it absolutely required one key ingredient, what he called "urine earth." Howard did his research in British India on a big farm. He was in complete charge of all farming operations. It was 1930. Howard's farm was powered by oxen and the hoes and shovels wielded by humans — a great many humans who, as a result of British policy over the previous century had been made desperately eager to work as hard as they could for next to nothing. Howard's cattle were kept in a loafing pen. Once a year, just before the season of heavy rains, the pen was dug out six inches deep, and all this soil was heaped up in the composting yard, ready to be mixed into forming compost piles. The missing soil was replaced with topsoil from the farm fields. A backyard gardener with a few rabbit hutches could do much the same thing.

I suggest that when importing your annual dose of garden nitrogen, make it easy on yourself and get a guaranteed good result: use a potent organic substance such as oilseedmeal, feathermeal or fishmeal. These finely ground materials are still rough and irregular enough that mineral fertilizers, which are usually denser than seedmeal, tend to fall into little pockets and cavities in the seedmeals, allowing them to get well blended and keeping the blend from separating out if it should be vibrated or shaken. Whenever I am concocting fertilizers, I put the seedmeal component into the bucket first and then stir the other ingredients into the seedmeal. Works great!

Oilseedmeal

In my opinion, oilseedmeals are the ideal nitrogen fertilizer for food gardening. Oilseedmeals release slowly, but not that slowly. They are plentiful, relatively easy to buy, and inexpensive. They are manufactured as a byproduct whenever vegetable oil is extracted from seeds. This residue is a valuable feed for livestock and is highly useful as fertilizer. It is rich food, often given to dairy cows. Because it is edible, when oilseedmeal is scattered atop the ground, microanimals emerge from the soil to eat it during hours of darkness. They return to the soil during the day and release what remains of the now-digested seedmeal amongst the plant's roots. Try it! Sprinkle some seedmeal on the garden, keep the surface dampish at night, and watch it disappear over the course of a few days. Then sit back and watch your plants — now being nourished by rapidly decomposing microanimal poops — leap up and grow fast.

Purists sometimes oppose using oilseedmeal unless it is organically grown. That's fine, if you can afford organic (or if you seek certification). But in general, I disagree. Ours is now a toxic planet. Everywhere. I consider it impossible to avoid bringing contamination into the garden. I reckon most gardens receive many times more pollutants from the air, in the form of automobile and industrial exhausts, than from what might be in conventionally grown seedmeals, including GM seedmeals.

I believe seedmeal objectors are putting the cart before the horse. I, too, wish to create garden soil that independently produces its own nitrates. Yes, the sustainable ideal is to import nothing. But I know there are steps my soil must go through before it arrives in a condition where no nitrogen imports is a real possibility. Meanwhile, there is seedmeal. And meanwhile, for those concerned about agricultural contaminants in their seedmeal, there is coprameal, which usually is as close to an organically grown material as something uncertified can be.

Lately, another whole class of objections has arisen to fertilizing with oilseedmeals: the use of genetically modified varieties. The destructive goals behind the push for most genetic modification technology provides excellent reasons for an ethical person to not support it, even to the extent of refusing to use GM waste products. On the other hand, my intuition tells me that when it comes to seedmeal decomposing in the earth, seedmeal is seedmeal. Even if some GM proteins are a bit kinky, the soil bacteria will make do. What about the fact that GM seedmeals carry traces of glyphosate and assorted pesticides? Well, conventionally grown oilseedmeals carry traces of different herbicides and pesticides. In fact, part of the excuse behind developing Round-Up-resistant oilseed plants was to avoid having to use even worse herbicides.

What we're mostly talking about here concerns making an ethical choice, not a scientific one. Do I use an inexpensive, effective nitrogen fertilizer — perhaps the best natural garden fertilizer value there is right now — or do I avoid this material in order to ever-so-slightly inconvenience corporations pushing genetically modified cotton, canola, soy, etc.?

Making ethical choices is not easy. Most people prefer to operate on the level of morality rather than get a headache over ethics. Morals are the easy way to go through life. You get handed a list of commandments. Obedience to them is good. Other behaviors are bad, and that's that. And having made the straightforward moral decision, if a situation does not work out as you'd hoped, rest easy, it wasn't your fault; you did the right thing. But if you sort out a perplexing situation by asking yourself to determine, on the basis of your own limited experience and flawed wisdom, which of the many possible choices results in the greatest good for the greatest number — the definition of ethics — and if your actions do not create the result you hoped for, then it is you who are to blame; it was your choices that led to that result.

A friend of mine in Maryland who struggled with this decision about seedmeal, came down on the GM-is-acceptable side. He has a big garden (one acre). His nitrogen mostly comes from GM soybean meal and has for years. He makes no effort to find non-GM seedmeals. He uses a huge worm bed to compost much of his crop waste. To keep his worms fed during winter/spring, when there is never enough fresh material for them, he feeds them soybean meal. This practice has been going on for nearly five years now. His worms seem happy. Of course, straight soybean meal is only a supplement to their mainstay, which is garden wastes grown with soybean meal. And, as I joked with my friend, the only way to know for sure if this seedmeal is really, really safe would be to feed worms exclusively on soybean meal and then feed those worms up the food chain, say to frogs for three or four of their generations, and then see how the frogs are doing.

In accord with the ongoing depletion of our farm soils, today's oilseedmeals are not as potent as they were in the early 1980s. In those days, seedmeals delivered more P (3%–4%) then they presently do (2%–3%). I ascribe that to ongoing depletion of industrial farm soils, and it goes hand in hand with the overall reduction in nutritional values amongst all other industrial foods being fed to humans. I recall oilseedmeal protein levels being a few percentage points higher 30 years ago, too. When working out a remineralization program, I ignore any phosphorus or potassium content in oilseedmeals. Basically, I am buying slow-release nitrogen at the best possible price; any minor phosphorus or potassium content is usually welcome but goes uncounted.

I mentioned earlier that temperature determines the rate of nitrogen release. The speed of decomposition also depends on the concentration of nitrogen in the material being decomposed. Microorganisms that eat organic materials must first build their own proteins from nitrogen being released by the materials they are digesting. If organic matter carries insufficient nitrogen to rapidly build a ravenous microbial population, then its breakdown happens too slowly to create a strong growth response. In my experience, if an organic fertilizer provides less than 4% nitrogen, it doesn't act as a strong fertilizer unless the soil is quite warm; using low-potency organic nitrate fertilizers in spring won't make you smile. Rapid release also requires a high soil-air supply. This explains to me why some people have excellent results using lower-potency materials like alfalfa meal or used coffee grounds and others do not.

Finally, if you are offered some other sort of less commonly found oilseedmeal, like sunflower, flax (linseed), sesame, safflower or peanut, don't hesitate. Simply check the label for protein content, divide by 6.2 (or multiply by 1.6, same difference), and there's your nitrogen content. Then compare the price per unit of nitrogen with other seedmeals, and choose the best value. Sometimes farm supply merchants don't stock processed seedmeal, but do offer ground oil seed, which is sold as animal feed. This meal has not had the oil extracted from it, so its protein percentage is reduced by the amount of that oil — and worse, it usually ends up being quite a bit more costly. The oil content will not harm anything in the soil, but its lowered effectiveness is not worth the higher price. If offered that stuff, shop harder.

Give some thought to the safe storage of seedmeals. They're edible; they'll interest mice and rats. Blending them with lime and phosphates (such as when making the Complete Organic Fertilizer I describe in Chapter 4) greatly reduces their appeal to vermin. If properly dry seedmeal is stored in metal garbage cans or oil drums with tight lids, it'll keep for many years. Beware: I once stocked up with an extra year's supply of local Tasmanian canolaseed meal that had not been properly dried; it slowly formed mold even though it was sealed inside a 44-gallon steel drum with a vermin-tight lid.

Material	N%	P%	K%
Alfalfa meal	2–3	1	2
Cottonseed meal	6–7	0.4–2.0	1.5–2.0
Soybean meal	7	2	1
Feathermeal	7–12	0	0
Fishmeal	8–10	4–12	0–2
Coprameal	3.5–4	1	—

Table 6.1: Nutrient Content of Seedmeals.

Fishmeal

As fertilizer, fishmeal stands head and shoulders above oilseedmeals. It contains nearly twice the amount of nitrogen usually found in oilseedmeal. It contains meaningful amounts of phosphorus, whereas oilseedmeal often does not. Most significantly, coming from the ocean, it contains every micronutrient; in this respect, it is somewhat like kelp meal. If I include the value of the phosphorus in the fishmeal (using the price of P as found in SRP), and allow for its higher concentration of nitrogen, fishmeal works out to be about equal in cost to oilseedmeals.

There is one practical liability to using fishmeal — the odor. Pets and wildlife will find it irresistible for a few days, and humans usually find it disgusting. It smells like cheap, tinned cat food — times ten. Store it in a tight metal container or in a feed bag hanging from a wire, so critters can't get into it.

There are ethical considerations about fishmeal, too. Fishmeal is ground-up fish. Its use depletes the ocean of feedstocks for larger, more economically desirable fish. And it comes from an increasingly polluted ocean. One thing to be careful of here: fishmeal is often fed to farmed fish in the form of pellets. Sometimes these feeds are made exclusively from oceanic fishmeal and sometimes they contain considerable amounts of (GM) soy protein.

Feathermeal

A byproduct of industrial chicken-raising, feathers are minimally processed for use as cattle feed. As fertilizer, feathermeal must be incorporated into moist soil to release its nitrates. This very slow-release nitrate product is highly desirable in gardens. It is usually about double the nitrate potency of oilseedmeal, but contains little or no phosphorus or potassium. I suggest that you use oilseedmeal in springtime. If the soil is warm, feathermeal will work great. In most circumstances, it might be wise to use half and half seedmeal/feathermeal.

Coprameal

After oil has been extracted from dried coconut meat (copra), this pleasant-smelling meal is on the low-end of providing sufficient potency for use as fertilizer. Coprameal is mainly used to feed racehorses. One big plus of coprameal is its purity: coconut trees are not chemically fertilized (they are almost impossible to spray). The coconut tree has a special ability to access otherwise unavailable minerals — which are still present in the meal. Copra is a developing-world, low-tech product; using coprameal will help support a great many struggling rural people.

Compost

To get a strong growth response from compost it must contain more than 3% N *and* the carbon-to-nitrogen ratio must be 12:1, or better 10:1. Most home-made compost is not this effective. Many brands of bagged compost are not that effective either; read the label. By the way, nitrogen percentage printed on bagged compost or on a lab analyses is calculated on a dry weight basis. If you've got a ton of moist compost, its dry weight might be half a ton. If you want to spread 100 pounds of N per acre, and your compost is 3% N, you might think 3,300 pounds (a ton and a half) of that moist compost contains 100 pounds of N. Actually, 6,000 pounds of moist compost might have a dry weight of 3,300 pounds. But it is even worse: this is compost, not fertilizer, so not all that nitrogen will be released in one crop cycle. Maybe half of it will be — maybe. So, to be confident of gaining 100 pounds of N per acre, you have to spread 12,000 pounds (6 tons per acre). How much is that in real life? It comes to a thin covering one-quarter-inch thick.

Manure

It is poor practice to side-dress with fresh manure as a nitrate fertilizer. As a mulch, it will generally provoke some growth, but it also releases undesirable breakdown products that can unappetizingly flavor your crop. And, lying on the soil's surface, especially if in the sun, a lot of the nitrates in raw manure will off-gas as ammonia. In the same way that the protein and phosphate contents of oilseedmeals have declined in the last decades, human bodies have on average become more fragile; these days many people risk illness from contact with raw animal manure. When our overall food supply contained more nutrition, this was not the case.

Composted chicken manure, sold inexpensively in sacks, does produce a strong growth response, if the analysis exceeds 3% (it is not uncommon to see it at 3.5% and even 4%). I have side-dressed crops with bagged chicken manure compost and enjoyed a strong growth response.

Legume Seedmeal

A Tasmania seed company specializing in pasture/forage crops is deeply interested in the possibilities of growing the wild white lupin (*Lupinus alba*) for use as nitrate fertilizer. Per acre, this species yields more tons of seed containing more protein than oilseed crops do. Wild white lupin seed contains about 20% (inedible, very bitter, maybe poisonous) oil that can be made into biodiesel; extracting that oil crushes the seed, which can then be used as fertilizer. The white and blue lupin are uniquely capable of uptaking phosphorus from subsoil reserves of highly insoluble iron phosphate, something almost no other crop can do. If sold for fertilizer after the oil has been extracted, the crop is highly profitable at the usual bulk price of oilseedmeal. Wild lupin seeds contain an extremely bitter (and toxic) alkaloid. While the seeds are decomposing, the alkaloid suppresses soil disease organisms, thus bringing more desirable microorganisms into a dominant position in the soil ecology. I think the lupin seed's effect on the soil ecology is the reason my own garden vegetables respond more strongly to being fertilized with ground lupin seedmeal than from getting the same quantity of N by way of oilseedmeal.

Therein lies a hint for the frugal. Anytime you can obtain low-germ or non-germinating legume seed and can grind it even into coarse chunks, you've got some excellent garden fertilizer. Legume seeds vary in protein content. The lowest are garden peas and the ordinary garden bean, *Phaseolus vulgaris*. The highest are lupins and fava beans, especially the small-seeded fava varieties. I've never tried fertilizing with flour made from non-germinating clover seeds, but I bet it would work.

The "Trace" Nutrients

Until such time as we can purchase rock dusts that contain concentrated levels of zinc, copper and manganese, we are going to have to use sulfate fertilizers. This reality has been recognized by the organic-certification bureaucracies; sulfates are now allowed. Sulfates do not damage the soil ecology or the soil itself. Yes, their manufacture consumes non-renewable energy, but choosing not to use them for this reason makes for a much poorer nutritional outcome.

Sellers of rockdust suggest that you can meaningfully increase minor-nutrient levels by using ordinary rock dusts. Many garden writers uncritically repeat these assertions. To verify my gut-feeling rejection, I crunched the numbers on several sorts of rock dust for sale. A typical analysis given in ppm looks impressive; every plant nutrient you could possibly want is on the list, but when these numbers are made real, the result isn't at all positive. What do I mean by "made real"? Well, when working with soil analysis, 1 ppm = 2 lb/ac in the two million pounds of a furrowslice acre. So, I computed the amounts of a few elements that might be present in 20 tons of highly mineralized basalt rock dust. Twenty tons per acre provides 1 pound elemental zinc per acre, 4 pounds of copper, 2 of boron, 66 pounds of phosphorus and 50 pounds of sulfur. The amounts of sulfur, phosphorus, copper and boron could be meaningful if the rock dust dissolved rapidly in soil — but, normally it takes many years to dissolve, no matter how fine the grind. One pound of zinc is not enough to improve much of anything, even if it were instantly soluble — which it is not. And consider: that's at 20 tons per acre! A store called Concentrates in Portland, Oregon, sells basalt dust at $880/ton in sacks. If we could purchase finely ground basalt dust at prices similar to the cost of ag lime, say $20/ton including the cost of spreading, then using it might make sense on a farm field. Need I say more?

What we really need to use are finely ground mineral ores of the sort sent to refineries to make pure zinc or copper or manganese. Meanwhile, there are the sulfates.

Buying Sulfates

Tasmanian farmers routinely buy sulfate compound fertilizers in 55-pound bags. But a full sack of zinc or copper sulfate might be sufficient to supply a big garden for a few decades — or a lifetime. Fifty pounds of zinc, copper or manganese sulfate might supply a neighborhood soil analyst for a year or two. An investment in one bag of each of these sulfates (including iron and potassium) would cost only a few hundred dollars. But, if you cannot find anyone in your area willing to sell to you by the pound, and you are not ready, willing, or able to become an active soil analyst yourself, I suggest that you contact Black Lake Organics in Olympia, Washington. These folks will be happy to weigh out a small amount for you and send it by post or UPS. In Australia, sulfates are normally available at garden centers in half-kilo boxes. I anticipate it will be that way in North America, too, as more gardeners are awakened to the possibilities of using them.

Iron sulfate ($FeSO_4$)

There are two forms of iron sulfate; ferrous and ferric. The ferrous form is the one to use in gardening; it is pinkish or greenish. Ferric sulfate is a rusty-red color; it indelibly stains concrete and should be avoided. Both types of iron sulfate contain 30% iron and 18% sulfur.

Manganese sulfate ($MnSO_4$)

I know of nothing to caution you about regarding this substance. Manganese sulfate contains 32% manganese and 19% sulfur.

Copper sulfate ($CuSO_4$)

This material can be poisonous if ingested in large quantities. It is readily absorbed through the skin. It is dusty; within moments of opening a bag of copper sulfate, I faintly taste copper. Considering the legalistic culture of North America, I must warn you here to wear gloves and a face mask and provide good ventilation if handling the stuff. I am not personally so worried about copper. I think if a bit of exposure happens only occasionally, and if I take a bit of care to make sure there's good ventilation when I am briefly exposed to it, I consider that my body has merely picked up a bit of valuable copper nutrition. Were I exposed to it for hours or days at a time, I'd wear a mask. The blue form of copper sulfate (hydrated) contains 25% copper and 12.5% sulfur.

Zinc sulfate ($ZnSO_4$)

$ZnSO_4$ rapidly picks up moisture from the atmosphere and turns itself into a sloppy, wet mess. Store unused zinc sulfate in an air-tight container. It contains 35% zinc and 17% sulfur.

Foliar Feeding

A low-tech way to discover if plants are short a trace element is to foliar feed that substance once and see if you rapidly get a positive response. If the element is not needed, you get no response — but no damage done. There is a limit to how concentrated a foliar spray solution can be, so you can only effectively foliar feed elements the plant does not use in large quantity. Theoretically, you could spray NPK on the leaves. Once, I did experimentally attempt to almost completely supply a few large plants' need for N, P and K (and trace nutrients) with foliar feeding; to keep them growing fast, I had to spray three times a week.

Foliar feeding can have powerful, rapid results. Not long ago, a visiting agronomist pointed out to me that my Cucurbits were showing signs of zinc deficiency — leaf margins rolling over, leaves not completely filling out. They had been growing slowly for a while (although the visitor couldn't see that). I mentioned I had fed that soil 28 pounds of elemental zinc a few months previously, and at the time, the soil test indicated the garden was deficient by more than twice that much. The agronomist pointed out that my red soil was rich in iron, and all that iron interfered with zinc uptake. Next day, I foliar fed my Cucurbits a dose of zinc sulfate. Two days later, the vines were growing fast, and the new leaves looked normal.

About ten days later, my Cucurbits again stopped growing, and their leaf margins again curled. They clearly had run out of zinc. At this point, my garden was entering its final weeks of warmish weather (my early March is like the second week of September in Oregon). For many years, it had seemed normal that my zucchini and cucumbers started coming down with powdery mildew about the second week of March and often, by the third week of March, they would be falling apart. I sprayed zinc sulfate again anyway, this time at twice the previous concentration. Not only did the plants resume growth a second time, but the incipient powdery mildew vanished — not to reappear until early April! The vines continued to make rampant growth through the entire month of March, something I've never before seen on this property. I realize now that powdery mildew on my Cucurbits — in *my* soil and circumstances — arrived earlier than it otherwise might have because my plants ran short of zinc. That's not to say it will happen with your Cucurbits.

When you dissolve any salt in water and spray it on leaves, be it table salt or zinc sulfate, there is a "do not exceed" concentration beyond which damage may be caused. For most salts, one tablespoon per gallon is a comfortable and effective concentration — with the exception of copper sulfate. Copper makes a highly alkaline solution; it must be mixed about half the strength of other sulfates — about one heaping teaspoonful per gallon. You can dissolve more than one sulfate salt in the same spray tank as long as you do not exceed a combined one tablespoonful per gallon. At two tablespoons per gallon, some of the plants may get scorched, especially if there's a good amount of copper sulfate in the solution.

Measurement Equivalents

¼ teaspoonful = 1.25 gram

4 x ¼ teaspoonful = 1 teaspoonful = 5 grams

3 x 1 teaspoonful = 1 tablespoonful = 15 grams

2 x 1 tablespoonful = 1 ounce = 30 grams

Although the small quantities involved make boron appear to be a good candidate for foliar feeding, it is not. To be effective, boron must be taken in through the roots. If you think your plants lack boron, mix it to distribute at two pounds per acre, and spray it on the soil. If some of it gets on the leaves, no harm done. Just rinse it off soon after.

The weight of trace elements that can be safely added to soil in a single year may not be sufficient to completely vanquish a deficiency; sometimes it takes repeated applications over several years before levels build up high enough. Like my zinc-deficient cucumbers, plants may grow well for a few months after the element was mixed into the soil, but then they get into trouble. So it is wise practice to have on hand a bit extra of these sulfates; just a few ounces of each is enough. After you get good results with an application of copper, zinc, iron or manganese to the soil, if the plants later develop disease symptoms or stop growing rapidly, your first action should be to foliar feed them a dose of whatever element or elements you previously fed the soil. If this element has again become deficient, growth will pick up immediately, and disease manifestations may vanish almost overnight.

There's more information about foliar feeding in Chapter 8, where calcareous soils are discussed.

Micronutrients

These elements are required nutrition for plants and humans alike — in very small quantities. Expect them to be present in your soil. Some will be found at a pound or so per acre; some should be there in grams per acre. The standard soil test does not report on these elements at the standard price, but a more expensive test is available that will. Rarely is there any call for a gardener to test for these elements — a farmer may test for one in a region where there commonly is a micronutrient deficiency, but the gardener, no need. The gardener can afford to routinely include a full measure of one of several soil amendments that supply a broad range of these elements. It is also wise practice to alternate micronutrient sources. One year use kelp meal; the next, Azomite.

The usual suggested application rate for micronutrient amendments is one pound per 100 square feet per year. One pint of kelp meal weights about one pound. I recommend spreading double that amount per 100 square feet. There also are a full range of micronutrients in natural sea salt and in fishmeal. Greensand has many, too. Rock dusts are highly variable in respect to their micronutrient contents; do not count on them unless you have an analysis in hand and have worked out for yourself what the ppm (or parts per billion) on the analysis actually mean in practice. Hard rock phosphate also carries a range of trace elements (including some you might prefer it did not have).

Micronutrients include chromium, cobalt, iodine, molybdenum, selenium, tin, vanadium, nickel and fluorine.

Chapter 7

The Soil Prescription

Now it is time to address the other side of Matthew's worksheet. Our task is to select fertilizers to fill the deficiencies and calculate how much of them to spread (and buy). First, fill in your photocopy of the *Acid Soil Worksheet*'s left-hand column on page two, transferring any deficit amounts from the right-hand column on page one.

I apologize in advance that I cannot make the rest of the procedure into an easily-followed step-by-step. That's because natural fertilizers often contain more than one element — often, more of a secondary element than you want. On some soils, it takes a bit of juggling to work out the best compromises. And sometimes, in order to avoid creating excesses, you simply cannot put in all the minerals you want to — this year. But there's no need to stress: there's always next year.

There are some general principles you can apply to help solve your own soil puzzle. First, if any existing level is within 10% of your target, above or below, declare that "good enough." Ten percent is within the degree of accuracy you need to work with. Second, if you're short of funds, please do not skimp — no half-measures. Instead, fully remineralize as much square footage as you can afford to. Get a great result on a portion of your land instead of a mediocre result overall. Besides, I am confident that once you see the results, in the future you'll be eager to budget enough money for all the fertilizers called for, whatever it takes. Last, respect the application limits. A lot of experience went into creating them. The limits are your main protection against making serious errors. They allow adding enough to make a real difference and slowly build soil mineral reserve levels but prevent overdosing.

When you worked out the target levels on the front page of the worksheet, you first found the numbers for calcium, then magnesium, then potassium, which gave you the keys to work out the remainder of the elements. But solving the flip side of the worksheet goes differently. Expect to find yourself making false starts and then needing to begin again. To assist you with this, I have drawn some narrow columns on the right-hand side to add up the quantities of S, Mg, or Ca that may be coming along with some other element.

It usually works best to start with the sulfate salts of the minor nutrients — iron sulfate ($FeSO_4$), manganese sulfate ($MnSO_4$), copper sulfate ($CuSO_4$) and zinc sulfate ($ZnSO_4$). Then consider potassium, which is usually supplied as sulfate as well. Don't forget that each sulfate carries a percentage of the actual element you want *plus* a percentage of elemental sulfur. To make it quick and easy for you, the worksheet provides a fertilizer composition table. Where compound anions are involved (SO_4, P_2O_5), the percentage is given as only the P or only the S. Suppose you're looking at a 15 pound copper deficit; there is an application limit for copper of 7 pounds per acre. Copper sulfate is 25% copper, so it takes 28 pounds of copper sulfate to supply 7 pounds of elemental copper; that 28 pounds carries with it 12.5% sulfur, or 3.5 pounds. Pencil in "$CuSO_4$ — 28 lbs" in the wide central column and (the rounded-off) number "4" into the narrow "S" column in the horizontal line for copper.

There is a limit to the quantity of sulfur you should add in any one year. If the soil does not call for as much sulfur as potassium and/or trace elements bring with them, you have to make a choice: either reduce the amount of these sulfate fertilizers or else exceed your sulfur limit. If facing this choice, I'd first limit new potassium to 100 lb/ac, which is a reasonable quantity often used in farming and then, if necessary, I'd exceed the sulfur target by an extra 50 pounds. But please, never exceed your sulfur limit by more than that. (When we get to it, you'll see that the *Excess Cations Worksheet* does limit potassium to additions of 100 pounds/acre. But we're not there yet.)

Please don't stress about these choices. For perspective, keep in mind that before soil balancing, organic gardeners paid no attention to these minor nutrients and somehow managed to grow food — just less nutrient-dense food that had more disease problems than it might have had. And it'll be better next year, your soil's exchange capacity should be more balanced, needing less inputs overall; your fertilizer bill should decline.

Some people are good at working out puzzles like these. Others just go blank. If blank describes you right now, and if you can't get your partner or a friend to work the prescription out for you, I offer three excellent solutions: 1) There is an online web app that'll do the entire computation for you in an instant. Information on access to the program is provided in the Appendix, and Figure 7.2 at the end of this chapter is a reproduction of the spreadsheet at work on Matthew's soil analysis. 2) You can find a soil consultant to help you (the Appendix also points you to associations of neighborhood garden soil analysts). 3) You could use COF as discussed in Chapter 4. Once you've gone to the trouble of ordering a soil test, you know if your garden needs boron, manganese, copper or zinc. So you also know how much to increase the amounts of these elements in the COF formula. So you should be confident about making minor adjustments to your COF. In the same way, if your soil audit shows an abundance (or excess) of phosphorus or potassium, you could eliminate SRP or the potassium sulfate from your own COF.

I selected this example because it was a difficult exercise, one that required me to do a bit of mental juggling to arrive at my targets. It proved possible to give Matthew's soil everything called for and still keep within the sulfur deficit. Here's how I went about working it out:

The most important thing is to balance the four major cations. It is okay to supply only a fraction of the other elements, but you have to get the big four in balance if at all possible (without exceeding the application limit for calcium or magnesium). Matthew's new garden had a large potassium deficit, and, because all immediately useful forms of potassium come with a sulfur component, I first toted up how much sulfur would come along with the trace elements I wanted to add. For a first approximation, I worked out that it takes 263 pounds of manganese sulfate to provide 84 pounds of actual manganese; 263 pounds of $MnSO_4$ would bring with it 50 pounds S. Another

	Deficit From other side of worksheet	Application Limit Per acre/year	Quantity and Material to Add	S	Mg	Ca
Sulfur **S**	*135*	110 lb 90% Ag Sulfur		*133.5*		
Phosphorus **P**	*Excess*	175 lb/ac elemental P				
Calcium **Ca**	*244*		*Ag lime 451 lb/ac*			*176*
Magnesium **Mg**	*164*	No more than 10% of target magnesium per year	*Dolomite lime 284 lb/ac*			*62*
Potassium **K**	*225*	200 lb/ac elemental K	*Potassium sulfate 476 lb/ac*	*80*		
Sodium **Na**	*47*		*Sea salt 135 lb/ac*			
Boron **B**	*2.7*	2 lb/ac elemental B	*Borax 20 lb/ac*			
Iron **Fe**	*Excess*					
Manganese **Mn**	*84*		*Manganese sulfate 263 lb/ac*	*50*		
Copper **Cu**	*8*	No more than 7 lb elemental Cu	*Copper sulfate 28 lb/ac*	*3.5*		
Zinc **Zn**	*Excess*	No more than 14 lb elemental Zn				

	N	P	K	S	Ca	Mg
Fish Bone	4	8.8		.06	19.0	.03
Fish Meal	10	2		0.6	2.3	.03
Crab Shell	3	1.5	.025	.02	23.0	1.3
Blood Meal	13	0.5				
Feather Meal	12	0.0	0.35	0.4	0.6	
Bone Meal****	3	13.0		2.5	12.0	0.3
Oilseed Meal	6	1.5	1.0			
Copra Meal	4	1	0.7			
Kelp Meal	1	0.3	2.5	2	2	0.7
Ag Lime					32-39	2
Dolomite					22	13
Gypsum				17	20.5	
Oyster Shell					36	0.03
Magnesium Oxide						50
Montana Hard Rock Phos**		1.3			29	
Calphos		8.8			20	
Monoammonium Phosphate		23 (Plus 12% N as NH_3)				
K-Mag			18.2	22		11
Langbeinite			15.6	23		12
Greensand***		.05	6	1.3	1.5-3.0	2-4
Ag Sulphur				90		

Sea Salt	35% Sodium (Na)	
Borax	10% Boron (B)	
Iron Sulfate	18% S	30% Fe
Manganese sulfate	19% S	32% Mn
Copper Sulfate	12.5% S	25% Cu
Zinc Sulfate*	17% S	35% Zn
Potassium Sulfate	17% S	42% K
Magnesium Sulfate	13% S	10% Mg

* Zinc sulfate picks up moisture from the air; store in airtight container.

** Hard Rock Phosphate is 1.5% available P and contains around 27% insoluble phosphate.

*** Greensand contains 9% Fe, 50% Si and many trace elements. More than half its potassium content is insoluble.

**** Bonemeal contains 5.7% sodium.

Fig. 7.1.

4 pounds S comes from the copper sulfate. So, so far, I have 54 lbs of sulfur.

Matthew's magnesium level is short by 164 pounds; his target is 374 pounds. Ten percent of that target, 37 pounds, is the application limit in this case. If I use Epsom salts to supply 37 pounds of magnesium, I have too much sulfur — and a bigger fertilizer bill. If I call for dolomite lime to supply the 37 lb/ac magnesium requirement, I would need 284 lb/ac, and, along with the magnesium comes 68 lb/ac calcium. The remainder of the calcium requirement is satisfied by using ag lime. (Regarding magnesium, I have some opinions: For one, too much magnesium causes soil compaction. For another, I've read that plants rarely suffer shortages of magnesium at the tissue level, even when the topsoil is a bit deficient. If there is one major cation I am willing to leave for next year, it is magnesium.)

Back to potassium. If I use potassium sulfate to supply the 200 pound application limit, I would need 476 lbs K_2SO_4, which brings with it 80 lbs S. That much S if combined with the amount of sulfur in the full requirement for manganese sulfate puts me just over the sulfur limit.

Next year's test should not call for nearly as much potassium; the manganese shortfall should be much less next year. However, it could

Application Limits

To avoid shocking soil or creating damaging excesses if a computation error is made, the *Acid Soil Worksheet* suggests limits on the amounts of fertilizers applied at one go. They are:

- Sulfur: 110 pounds ag sulfur
- Phosphorus: 175 lb/ac elemental P
- Magnesium: overdoses of Mg tighten soil; it can take many years for this to work out. Limit Mg applications to 10% of target level.
- Potassium: 200 lb/ac elemental potassium
- Boron: 2 lb/ac elemental boron
- Copper: no more than 7 lb/ac elemental copper
- Zinc: no more than 14 lb/ac elemental zinc

Conversions

1 lb/acre = 1 kg/ha = 1 gram/100 sq ft

1 lb/acre = 10 grams/1,000 sq ft = 1/3 ounce per 1,000 sq ft.

go the other way. If Matthew spreads compost, the TCEC should be higher next year, and the larger pantry he creates will demand more of everything.

There are no issues regarding the requirements for sodium and boron. Sea salt is about 35% sodium; it takes 135 pounds of agricultural salt to bring 47 pounds sodium. Borax is 10% elemental boron; 20 pounds of borax contains about two pounds boron, our application limit.

So here's the prescription, so far:

All quantities are in pounds per acre, which approximately equals grams per 100 square feet (So, 284 *pounds per acre* of dolomite translates to 284 *grams per 100 square feet.*)

- Agricultural lime: 451 pounds per acre
- Dolomite lime: 284 pounds per acre
- Potassium sulfate: 476 pounds per acre
- Sea salt: 135 pounds per acre
- Borax: 20 pounds per acre
- Manganese sulfate: 263 pounds per acre
- Copper sulfate: 28 pounds per acre.

The Standard Prescription

From the front page of the *Acid Soil Worksheet* (Figure 5.2), we see that the organic matter content of Matthew's soil is 5.4%. For a grassy field with soil barely above the light/heavy line, this amount of soil organic matter is not too bad. Since this garden is not seriously low in organic matter, its first dose of compost could be only a half-inch thick. Even if Matthew were to spread only a quarter-inch of compost, I predict the organic matter level next year will test over 6%, and the TCEC will be up.

Matthew's field also needs a source of micronutrients; my preference is kelp meal over Azomite — but that's my preference, there's no hard rule.

And then there's nitrogen to consider. Matthew's 5% organic matter could release 75 pounds N in the next year, mostly in high summer. But to grow superlatively, especially before summer's heat arrives, a veggie garden needs more N. And where Matthew is located, the phrase "hot summer" is an oxymoron. So, for early-season nitrogen, I usually call for seedmeal — three quarts seedmeal per 100 square feet is plenty (100 lbs N/acre).

So here's the full prescription (for each 100 square feet of garden):

Mix together uniformly:

Agricultural lime: 451 grams (use #100 grind if possible; if not, use #65)

Dolomite lime: 284 grams (Use #65 grind or finer.)

Potassium sulfate: 476 grams

Sea salt: 135 grams

Borax: 20 grams

Manganese sulfate: 263 grams

Copper sulfate: 28 grams

3 quarts oilseedmeal, or 2 quarts oilseedmeal and 1 quart feathermeal

1 quart kelp meal.

Blend all fertilizers uniformly. Over each 100 square feet of growing area, spread compost ½-inch thick; cover the compost uniformly with the fertilizer mix. Dig it all in six inches deep.

When I analyze a soil test for other people, I always ask them to retest at the same time the following year, so I can see what shifts occurred. If a soil has a deficiency of iron, manganese, zinc or copper, you, as a soil analyst, would tell your client this: If, during the growing season, any disease problems should develop, or growth slows down, the first remedy to try is one teaspoonful of a sulfate fertilizer (though only ½ teaspoon of copper) dissolved in one quart of water. Spray it on the plants until water drips off every leaf. If that remedies the situation in a few days, then start regularly foliar feeding with that element; next year, you should add more of that element to your soil.

Matthew

Test Date:	02/03/12	Report Date:	07/07/12

Logan Labs Soil Report Results		
Sample Location		Unfarmed
Sample ID		West
Lab Number		50
Sample Depth in Inches		6
Total Exchange Capacity (M.E.)		12.87
pH of Soil Sample		5.80
Organic Matter (%)		5.36
Sulfur:	ppm	26
Mehlich III Phosphorus	as (P_2O_5) lbs/acre	954
Calcium:	Desired value	3501
lbs/acre	Value Found	3292
	Deficit	-209
Magnesium:	Desired value	370
lbs/acre	Value Found	210
	Deficit	-160
Potassium:	Desired value	401
lbs/acre	Value Found	125
	Deficit	-276
Sodium:	lbs/acre	73
Calcium (60 to 70%)		63.93
Magnesium (10 to 20%)		6.8
Potassium (2 to 5%)		1.24
Sodium (.5 to 3%)		1.24
Other Bases (Variable)		5.8
Exchangable Hydrogen (10 to 15%)		21
Boron (ppm)		0.63
Iron (ppm)		206
Manganese (ppm)		8
Copper (ppm)		5.16
Zinc (ppm)		31
Aluminum (ppm)		949

Comments

See notes below left

Enter the area to amend and PRESS ENTER ------------→	1	acre(s)

Choose units of weight ----------------→	lbs / oz

Recommended Amendments		
Agricultural Lime	360	lbs
Dolomite Lime	285	lbs
Potassium Sulfate	457	lbs
Borax	20	lbs
Copper (Cu) Sulfate	28	lbs
Manganese (Mn) Sulfate	263	lbs
Total weight of all amendments	1412	lbs

Organic matter is good at over 5%. Deficient in sulfur, magnesium, potassium, boron, and manganese. Potassium Sulphate application is only 91% of target because of sulfur target. Excesses including iron and zinc are of minor importance. Sodium target set to a cautious 1% - irrigation water will provide more sodium. Lime application will take 3 years to become fully effective. Apply feathermeal at 20 lbs/1000 sq ft for nitrogen. If you enter the measured area of your garden, this spreadsheet will calculate the weights of all amendments for you.

☐ Check this box to override "Consult a Soil Analyst" warning, and view automatic recommendations. This should be used cautiously, since "Consult a Soil Analyst" warns of a situation beyond the capability of this spreadsheet.

Notes:

<> Amount per application limit was reached for these elements, compounds and/or amendments:

Mg, Cu, B,

<> Retest next year

Credits for the algorithms: William Albrecht, Michael Astera, Neal Kinsey, William McKibben, Gary Zimmer, Spectrum Analytics, Steve Solomon, Erica Reinheimer

Fig. 7.2: *Matthew's soil analysis done by the Reinheimer WebApp, OrganiCalc™, at GrowAbundant.com. Page 1 of 3.*

Chapter 8

Soils with Excesses

Excess pH

Simple logic: Soil with a pH over 7.0 often has too much of its exchange capacity loaded with magnesium, potassium, or sodium, instead of with calcium. There are two main questions to ask yourself concerning excess pH: Can the pH be lowered in a gentle manner consistent with being a steward to the land and environment? And: In order to grow an abundance of nutrient-dense food, do these excesses need to be dealt with at all?

The first point to settle concerns lowering pH when it exceeds 7.0. Can this be made to happen? And at what cost? At what risk? Is it even necessary?

Most food crops grow excellently with a soil pH as high as 8.2 (a few actually grow best around 8.2). As stated before, most crops do grow best when pH is 6.4, but many still do fine at 8.2. So, in and of itself, a soil pH ranging from 6.0 to 8.0 is a minor matter to most food crops. However, the availability of plant nutrients is strongly reduced by high soil pH. And this is vitally important.

Please take another long look at the crucially important chart of pH availability, reproduced again on the next page, for your convenience.

As soil pH moves from 7.0 to 8.0, most nutrients become far less available. For some, like sulfur and magnesium, the effect is not that severe. However, potassium loses about three-quarters of its availability between 7.0 and 8.0, and phosphorus does even worse. Iron, zinc and manganese squeeze down to nothing above 7.5. Obviously, if

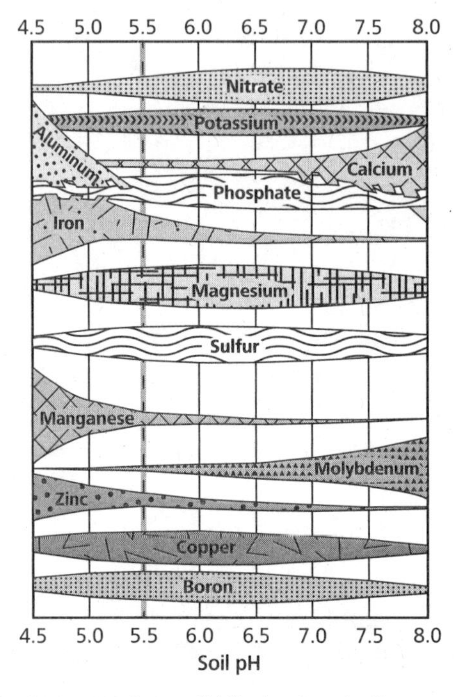

Fig. 8.1: Nutrient availability in mineral soil.

soil pH can conveniently be reduced below 7.5, the garden will be better nourished.

You will understand this better if you consider how soils develop — how they transform while going from youth to old age. Young soils normally contain a full complement of the four major cations, and there usually will be far more calcium than the other elements. As soil ages, its fragments of mineral-containing rock gradually dissolve into the soil solution. Some of these dissolved elements recombine to form clay; the rest get leached out. Given enough time, you'll have a clay soil holding only a tiny remnant of its original rock fragments. Soil scientists routinely measure the percentage of undecomposed rock fragments left in an old clay soil to evaluate what potential it still has to release more nutrients.

Clay declines in agronomic value as it ages. Young soils with some clay content can have a TCEC in the range of 30–40 in the top six inches. Over geologic time, the exchange capacity of clay drops, eventually to a very low level. The amount of reserve minerals (that have not yet dissolved) declines as well. So, old soils may be thought of as being weak or exhausted (the proper term is *developed*); young soils are, in a way, strong, or *undeveloped.*

Most of the soils in the northern United States (which was covered by a continental glacier 10,000 years ago) are still fairly young. Somewhat leached, but still young. Typically, they possess strong clay. Many hold large, unreleased mineral reserves; with proper management, these have every chance to remain naturally productive agricultural soils for thousands of years. Because of the nature of the rocks the most recent continental glacier brought from the Canadian Shield, soils in the northern United States tend to magnesium excess.

In the American Southeast, the soils are enormously older; they are in their end-game, having been leached of the greatest part of their original mineral endowment. Their acidic clays are weak. The TCEC of these soils is generally below 10. Sometimes 4.

Moving our attention westward, where the evapotranspiration ratio is below 80, you'll find highly mineralized, slightly leached, high-exchange-capacity prairie soils laden with plant nutrients, often with a pH between 7.0 and 7.6. Soils like this are found from Texas to Saskatchewan and Alberta. These have proven to be the most productive cereal-growing soils in North America.

Now consider the other extreme. In the semi-arid and desert regions of the southwestern United States, the soils are young and unleached because the initial stages of soil formation proceed rapidly in arid climates. With little vegetative cover, desert mountain slopes disintegrate from frost cracking, wind erosion and occasional heavy rainstorms (not uncommon in deserts). Thick deposits of fresh alluvium washed off the surrounding highlands fill in low-lying areas, forming broad, flat alluvial valleys, or plains. Where desert soils form more slowly out of solid rock, the minerals will not have been leached. In either case, alluvial or upland, arid soils tend to be highly mineralized. Usually the most prominent mineral is calcium, although it is sometimes magnesium. East of the Cascade Mountains, potassium often predominates. In some places, huge excesses of sodium plague agriculturalists.

Soils with excesses are common in arid and semi-arid regions in New Mexico, Arizona, Utah, Texas, Nevada and east of the Cascades. Sometimes there is so much calcium present that moisture-deposited lumps or layers of calcium carbonate are concentrated in the topsoil. Sometimes, the free lime is more dispersed. It depends on the quantity and frequency of rainfall, topography, soil texture and vegetation. Calcium can deposit as a whitish layer called "caliche," that prevents good drainage and is often (but not always) impervious to root penetration. Caliche forms where limited rainfall occurs at a nearly constant annual rate; the calcium is repeatedly leached to a particular depth where it is deposited as the soil dries out. Caliche can also be caused by irrigation.

Continuing this survey to the West Coast, the windward side of the Sierra Nevadas gets enough rainfall some winters to cause leaching. Some California agricultural soils are slightly acidic. Some are a bit over 7.0. Some have considerable excess magnesium; these are called "serpentine soils" (after the type of high-magnesium rock they formed from). Regions that have mostly calcareous soils are found in Texas, Iowa, Kentucky and Wisconsin (and elsewhere).

When analyzing soil test results, the critical dividing line happens around pH 7.6. If your soil pH is between 7.0 and 7.6, the *Excess Cations Worksheet* will certainly help you improve your situation. It will also help with soils with pH higher than 7.6 — as long as they are not calcareous. If the pH exceeds 7.6, and that high pH is caused by free calcium in the soil, I see little sense in trying to lower it. There's a section about handling calcareous soils at the end of this chapter and a special worksheet specifically for them in the Appendix.

Irrigation

Erica Reinheimer, who endlessly contends with excess magnesium and sodium, said, in a personal communication:

> *In arid soils, excesses are often the result of limited rainfall — there is not enough rain to flush minerals deep into the subsoil. These lands can be leached of excess minerals IF enough high-quality irrigation water is available. However, the aquifers in such regions often contain too many excess minerals to make this effective. If you are growing a garden in an arid climate, you need to know a lot about the minerals in your irrigation water. When you start to irrigate, the minerals in your water will become more important than the minerals you see in your soil report.*

Rainwater is quite pure, especially when the air contains little sulfur from coal power plants or diesel fuel. When it rains, the soil solution is diluted, but its nutrient balance is not altered. When irrigation commences, and if there are high concentrations of minerals in the water,

Sodium compounds are extremely soluble and are present in pretty much all irrigation water — usually at low levels. You should know what your annual sodium input from irrigation is before adding any salt to your soil. If you are on city water, your supplier issues an annual water quality report. You should also check boron, copper, and zinc levels in irrigation before adding these minerals.

these minerals can quickly overwhelm the soil solution. So knowing what minerals are already in your soil only reveals part of the picture.

To figure your annual application of these soluble minerals in pounds per acre, you first estimate your garden's annual water deficit as 1½ inches per week times the number of weeks you irrigate (but don't worry about the size of your garden). This will give you a number that will correspond to your acre-ft/year irrigation application. Suppose you're in California or Oregon, irrigating once a week from June through September, regularly, no rain anticipated, applying 1½ inches each time. Sixteen weeks times 1.5 inches per week would be 24 inches. If you were irrigating an entire acre this way, you would spread two acre-feet over the summer. One furrowslice acre of soil (which is six inches thick) has about the same weight (two million pounds) as the same volume of water. Therefore, 24 inches — two acre-feet — of water weighs about eight million pounds. If your water analysis shows it contains 50 ppm sodium, that would be 50 pounds of sodium in one million pounds, or 400 pounds of sodium in eight million pounds (the weight of two acre feet of water). That's a great deal! Good thing sodium is the easiest of all the major cations to kick off the exchange points. And good thing that in some winters it rains hard in California. Sodium is readily leached by rain water, especially in well-balanced soils. Hopefully, you will not see any accumulation on your soil report. Excess sodium is a real problem for your plants.

Irrigation water has pH. It can be quite high in some areas because of dissolved calcium and magnesium. This can explain why, especially on low-TCEC soils, the pH never seems to go down much, despite good mineralization practices. Regularly irrigated soils may have to be constantly fed gypsum in order to constantly leach the magnesium being added by irrigation. If mineral accumulation is happening, you'll see it on your soil audit. Please recall that the system calls for sulfur at half the usual magnesium target (when magnesium is targeted at 12% saturation) if there are any excesses. As long as you are irrigating in a circumstance like this one, sulfur should be kept at the higher level and should be applied as gypsum. It is possible that your water quality is so poor that more gypsum will be needed than this book's system allows. If that seems the case, I suggest you contact a soil analyst.

Excess Cations Worksheet

Many who use the nutrient-balancing approach come to believe there is an ideal soil. If their garden doesn't match that, or isn't heading towards that profile, they think there is something wrong. Best you do not invalidate your own garden with this kind of thinking. And who told you in the first place that there was only one ideal soil profile? William Albrecht worked out what would constitute a highly desirable range of saturation levels for the four major cations in the average acidic (non-calcareous) soil east of the Rocky Mountains. Albrecht's saturation targets also apply to Cascadia. If you ever get deeply into this topic, it'll help to keep in mind that the "average" soil in Albrecht's universe has a TCEC of 12.0, a pH of 6.3 and a potassium saturation of 4%. And, despite Albrecht's brilliance, it is quite possible he succumbed to the same malady many garden writers suffer from — succeeding in his backyard and expanding it to include the whole continent.

High-pH soils can be fertile and productive unless they have really big excesses — except for calcium. Excess calcium can be lived with. In fact, I am uncertain if it is possible to really have excess calcium. Cation for cation, magnesium lifts pH more than calcium. Sodium and potassium cations are even more effective at raising pH than magnesium is. Any of these three elements, in excess, can push pH over 7.0. If a magnesium-rich soil is light, and especially if it's sandy, being a bit overdosed with magnesium may not cause unwanted tightness. But, after seeing how my own clay loam soil loosened up after I brought the calcium-to-magnesium ratio closer to the proper balance for an acidic soil, I would have a hard time accepting excess magnesium if there were anything I could do about it.

Soil pH can unnaturally go over 7.0 if a farmer repeatedly adds fertilizer but does not add lime. Some fertilizers have an acid reaction; those with a huge sulfur component combine with soil calcium to form gypsum, which then leaches out. Chloride fertilizers combine with calcium to form highly soluble calcium chloride, which readily leaches out. Through receiving dozens of such applications, the soil is leached of calcium, and not only the topsoil gets leached. Many agricultural soils are now like that. They are devoid of calcium, yet have high pH because the space on the TCEC that had been filled by calcium is now filled with potassium. The remedy for these high-pH soils is agricultural lime and gypsum. Sometimes, quite a bit of lime. Often these soils do not perform well until their calcium-deficient subsoil has been remedied.

If too much dolomite was spread, if the irrigation water contains a lot of magnesium, and if the soil naturally contains a lot of magnesium, an excess appears on the soil audit. These excesses can be reduced by applying enough gypsum to meet the soil's sulfur target. Excess calcium will be reduced by adding agricultural sulfur instead of gypsum. When magnesium, potassium or sodium excesses are accompanied by a calcium saturation below 68% (which constitutes a calcium deficiency), even though the pH is well above 7.0, enough ordinary lime can be added to lift calcium saturation to 68%, and then gypsum (up to the sulfur limit) can be spread to reduce the excesses.

If you live in the American Southwest or in limestone country, there is a high probability that you have an enormous quantity of calcium in your soil. To avoid confusing the issue with an additional soil test, and to save money, you can easily check calcium levels before sending off for a soil test. Mehlich 3 extractions do not work properly when there is more free calcium in the soil than you'd get from a slight case of overliming, so as a preliminary, do a simple vinegar fizz test. If the soil fizzes, you have calcareous soil. Otherwise the usual Mehlich 3 standard soil test will serve. I'll explain more about the fizz test later in this chapter in the section on handling calcareous soils.

Figure 8.2 shows an ordinary home-garden soil audit from eastern Kansas. Typical of prairie soils, it is loaded with cations. There are excesses of calcium, magnesium and sodium (which has a 1% saturation target on this worksheet). A fundamental goal for this soil is to lower its pH and eliminate those excesses.

Soil Report

Job Name **Cynthia S** Date 2/24/2012

Company Cynthia S Submitted By

			Recalculated for a 6 inch sample depth			
Sample Location		Veggie				
Sample ID		Patch 2				
Lab Number		89				
Sample Depth in inches		7				
Total Exchange Capacity (M. E.)		19.13				
pH of Soil Sample		7.70				
Organic Matter, Percent		10.53				
ANIONS: **SULFUR:**	p.p.m.	17				
ANIONS: **Mehlich III Phosphorous:**	as (P_2O_5) lbs / acre	351	300			
EXCHANGEABLE CATIONS: **CALCIUM:** lbs / acre	Desired Value	6071	5203			
	Value Found	6920	5930			
	Deficit					
MAGNESIUM: lbs / acre	Desired Value	642	550			
	Value Found	854	732			
	Deficit					
POTASSIUM: lbs / acre	Desired Value	696	596			
	Value Found	315	270			
	Deficit	-381				
SODIUM:	lbs / acre	109	93			
BASE SATURATION %: Calcium (60 to 70%)		77.50				
Magnesium (10 to 20%)		15.94		Cynthia sampled 7 inches deep. I have recalculated these numbers for a 6 inch deep soil sample.		
Potassium (2 to 5%)		1.81				
Sodium (.5 to 3%)		1.06				
Other Bases (Variable)		3.70				
Exchangable Hydrogen (10 to 15%)		0.00				
TRACE ELEMENTS: Boron (p.p.m.)		0.9				
Iron (p.p.m.)		222				
Manganese (p.p.m.)		53				
Copper (p.p.m.)		2.25				
Zinc (p.p.m.)		18.68				
Aluminum (p.p.m.)		291				
OTHER						

Fig. 8.2. **Logan Labs, LLC**

The *Excess Cations Worksheet* shown in Figure 8.3 has slightly different targets and restrictions than the *Acid Soil Worksheet*. Filling in the front page is the same. The main difference is the high sulfur level

Excess Cations Worksheet
U.S. Measurements

Name Cynthia S.
Plot or Field Veggie Patch
Date of Test 2/24/2012

Sample Depth 6 inches All numbers on this worksheet assume a six inch sample depth

TCEC 19 — The Mehlich 3 extraction overstates TCEC when free lime is present and should not be used if *the soil reacts when doing the Fizz Test*

pH 7.7 — This worksheet is for M3 audits on non-calcareous soils with a pH of above 7.0. If pH is over 7.6, do a Fizz Test. High pH can be caused by high levels of calcium, magnesium and/or potassium and/or sodium.

Organic Matter % 10.5 — Organic matter levels exceeding 5% are extremely helpful. From normal soil organic matter decomposition, assume approximate release of N = 15–25 lb N per 1% OM. Varies with temperature, moisture and soil air supply. N = 0.22 x NO_3

Element	Actual Level	Calculating Target Level (Pounds per acre)	Target (Pounds per acre)	Deficit
Sulfur **S**	ppm 17 lb/ac 34	**S = ½ Mg (Target Level)** until cation excesses eliminated; then switch to the Acid Soil Worksheet	274	240
Phosphorus **P**	P_2O_5 300 P = 132	**P = K (Target Level)** Calculate using actual P, not phosphate. P = 0.44 x P_2O_5	420	288
Calcium **Ca**	lb/ac 5930	**TCEC x 400 x 0.68 = Target Level**	5168	Excess
Magnesium **Mg**	lb/ac 732	**TCEC x 240 x 0.12 = Target Level** If deficit less than half of Target Level, do not add Mg	547	Excess
Potassium **K**	lb/ac 270	**K is proportional to TCEC: see chart**	420	150
Sodium **Na**	lb/ac 93	**TCEC x 460 x 0.01 = Target Level** Be certain of good water quality before adding sodium	87	Excess
Boron **B**	ppm 0.9 lb/ac 1.8	**B = 2 lb/ac if CEC below 10** **= 4 lb/ac if CEC above 10**	Do not exceed 4 pounds 4	2.2
Iron **Fe**	ppm 222 lb/ac 444	**Fe = 100 lb/ac if CEC below 10** **= 150 lb/ac if CEC above 10**	150	Ok
Manganese **Mn**	ppm 53 lb/ac 106	**Mn = 55 lb/ac if CEC below 10** **= 100 lb/ac if CEC above 10**	100	Ok
Copper **Cu**	ppm 2.25 lb/ac 4.5	**Cu = ½ Zn (Target Level)**	21	17
Zinc **Zn**	ppm 18.68 lb/ac 37.4	**Zn = 1/10 P (Target Level)**	42	5

Potassium Target Levels

TCEC	Pounds	TCEC	Pounds	TCEC	Pounds
		16	390	28	493
		17	400	29	500
		18	410	30	507
7	255	19	420	31	511
8	270	20	435	32	515
9	290	21	443	33	519
10	310	22	451	34	523
11	320	23	459	35	527
12	335	24	463	36	531
13	350	25	475	37	535
14	365	26	481	38	539
15	380	27	487	39	543

If TCEC is lower than 7, use value for 7. If it is over 39, use value for 39.

	One acre, six inches deep weighs	One hectare, 80 mm deep weighs
1 meq Calcium	**400 lb**	**400 kg**
1 meq Magnesium	**240 lb**	**240 kg**
1 meq Potassium	**780 lb**	**780 kg**
1 meq Sodium	**460 lb**	**460 kg**

1 ppm = 1mg/kg = 2 pounds/acre = 2.24 kg/hectare

Date of Issue: 07/07/2012

Fig. 8.3.

required to deal with excesses, so on this worksheet, S = ½ Mg. Also, the sodium target is reduced to 1% of total saturation. There are bigger differences on the reverse side (shown in Figure 8.4).

Any high-pH soil is, by definition, already overloaded with cations; for that reason alone, additions of magnesium and potassium are restricted. As excesses resolve, soil pH comes down accordingly, and potassium becomes more available. And it is always wise to be cautious about adding magnesium because, as pH drops, magnesium may appear, seemingly from nowhere. Regarding phosphorus in alkaline soils: if you're broadcasting it, my advice is to use monoammonium phosphate; it'll perform better than soft rock phosphate. Do not forget

Application Limits and Restrictions

The *Excess Cations Worksheet* limits fertilizer applications more severely than the *Acid Soil Worksheet* does:

- Sulfur: 110 lb/ac agricultural sulfur (100 lb/ac elemental S)
- Phosphorus: 175 lb/ac. Use only monoammonium phosphate or soft rock phosphate.
- Calcium: If saturation is below 68%, use fine grind ag lime in sufficient quantity to bring it to 68%. If there is excess Mg, K or Na, use gypsum in addition to ag lime in order to fill any sulfur deficit. If the excess involves magnesium, potassium or sodium, use gypsum for satisfying the sulfur target in preference to elemental sulfur. Do not be concerned if added gypsum brings with it more calcium than your 68% saturation target.
- Magnesium: Do not consider magnesium to be excessive unless the saturation exceeds 12%. Do not add magnesium unless saturation is below 6%, and in that case, do not attempt to raise magnesium saturation above 6%. However, when calculating sulfur at half the magnesium target, reckon target magnesium at 12% saturation.
- Potassium: 100 lb/ac elemental potassium
- Boron: 2 lb/ac elemental boron
- Copper: 7 lb/ac elemental copper
- Zinc: 14 lb/ac elemental zinc.

that if you do use MAP, your nitrogen requirement will be significantly covered.

	Deficit From other side of worksheet	**Application Limit** Per acre/year	**Quantity and Material to Add**	**S**	**Mg**	**Ca**
Sulfur **S**	240	110 lb Ag Sulfur	Reduce excess calcium with Ag Sulfur Ag Sulfur 110 lb/ac	100		
Phosphorus **P**	288	175 lb/ac elemental P	Use Soft Rock Phosphate if you are not willing to use Monoammonium phosphate. MAP 760 lb/ac (90 lb N as NH_4)			
Calcium **Ca**	Excess	If below 68% calcium saturation use ag lime sufficient to 68%	If excess Mg, K or Na, use gypsum to satisfy any sulfur deficit. If excess Ca, do not use gypsum, use ag sulfur.			
Magnesium **Mg**	Excess	No more than 10% of target magnesium per year				
Potassium **K**	150	100 lb/ac elemental K	Potassium sulfate 240 lb/ac	40		
Sodium **Na**	Excess					
Boron **B**	2.2	2 lb/ac elemental B	Borax 20 lb/ac			
Iron **Fe**	ok					
Manganese **Mn**	ok					
Copper **Cu**	17	No more than 7 lb elemental Cu	Copper sulfate 28 lb/ac	4		
Zinc **Zn**	5	No more than 14 lb elemental Zn	Zinc sulfate 14 lb/ac	2		

	N	P	K	S	Ca	Mg
Fish Bone	4	8.8		.06	19.0	.03
Fish Meal	10	2		0.6	2.3	.03
Crab Shell	3	1.5	.025	.02	23.0	1.3
Blood Meal	13	0.5				
Feather Meal	12	0.0	0.35	0.4	0.6	
Bone Meal****	3	13.0		2.5	12.0	0.3
Oilseed Meal	6	1.5	1.0			
Copra Meal	4	1	0.7			
Kelp Meal	1	0.3	2.5	2	2	0.7
Ag Lime					32-39	2
Dolomite					22	13
Gypsum				17	20.5	
Oyster Shell					36	0.03
Magnesium Oxide						50
Montana Hard Rock Phos**		1.3			29	
Calphos		8.8			20	
Monoammonium Phosphate		23 (Plus 12% N as NH^3)				
K-Mag			18.2	22		11
Langbeinite			15.6	23		12
Greensand***		.05	6	1.3	1.5-3.0	2-4
Ag Sulphur				90		

Sea Salt	35% Sodium (Na)	
Borax	10% Boron (B)	
Iron Sulfate	18% S	30% Fe
Manganese sulfate	19% S	32% Mn
Copper Sulfate	12.5% S	25% Cu
Zinc Sulfate*	17% S	35% Zn
Potassium Sulfate	17% S	42% K
Magnesium Sulfate	13% S	10% Mg

* Zinc sulfate picks up moisture from the air; store in airtight container.

** Hard Rock Phosphate is 1.5% available P and contains around 27% insoluble phosphate.

*** Greensand contains 9% Fe, 50% Si and many trace elements. More than half its potassium content is insoluble.

**** Bonemeal contains 5.7% sodium.

Date of Issue: 07/07/2012

Fig. 8.4.

This soil could use more phosphorus than is wise to give it at one go. MAP at 760 lb/ac contains 91 pounds of ammonium nitrogen, plenty of N considering the soil's organic matter level. Next year's soil audit should show the excesses have lessened. I expect that the pH will drop a few tenths, and hopefully, the availability of copper, zinc, sulfur and phosphorus will have increased because of a pH shift, not to mention because of what we're adding.

The soil prescription so far calls for:

- Ag sulfur: 110 lb/Acre. Combined with the other sulfates, the total sulfur component amounts to 146 pounds/acre.
- MAP: 760 lb/acre
- Potassium sulfate: 240 lb/acre, the 100 lb/ac potassium limit
- Borax: 20 lb/acre
- Copper sulfate: 28 lb/acre
- Zinc sulfate: 14 lb/acre.

In the event the gardener had been unwilling to accept MAP, I would substitute soft rock phosphate at 2,000 lb/acre. To replace the nitrogen in the MAP, I would add about 3 quarts seedmeal or 1½ quarts feathermeal per 100 square feet. The usual soil prescription also calls for kelp meal at 1 quart per 100 square feet. The soil organic matter level was remarkably high for this hot-summer climate; no further additions were suggested for the current year.

Excess Soil Air

Sandy soil cannot hold much moisture. Coarse sand soils can be so loose that strong winds can uproot plants. They can hold too much air, such that the soil ecology eats soil organic matter too rapidly, making it near-impossible to get the humus level up. A soil like that benefits from being tightened up, made a bit less airy, firmer, and able to hold a bit more moisture. An extremely coarse sand might perform best with a ratio of 62% Ca:18% Mg. A coarse, sandy loam might be best at 65% Ca and 15% Mg. The *Acid Soil Worksheet* presets that ratio at 68:12, but you can pencil in other levels to suit yourself. Select one of two ratios: 65:15 or 62:18; either way, Ca + Mg = 80% saturation.

Establishing this ratio should be easy enough in a light acidic soil that has not yet been limed — simply do not add more calcium than would push its saturation over 62%, and load up the Mg.

Recall that calcium clings harder to clay than any of the other major cations. So, if the soil in question already has a calcium saturation exceeding your preferred target, it may take some years to leach that calcium out. What makes this particularly difficult to accomplish is that adding sulfur to leach calcium makes gypsum, which then leaches magnesium — which you're simultaneously trying to increase.

If the soil is quite acidic and still has a significant calcium requirement as well as needing magnesium, then dolomite lime can be used, at least up to the point that calcium has been saturated at 62% or 65%. But to increase magnesium beyond that point with OMRI-approved ingredients, you need Epsom salts. These are costly and only contain 10% magnesium. So, even though it is not approved for organic certification, this is a case for the use of magnesium oxide.

Excess Calcium

I have often encountered the term "overliming." The idea is that the point of adding lime is to raise pH to somewhere between 6.2 and 6.4, and if more lime than that is spread, the amount is excessive. I have also read, not only in home-garden books, but in those written by practicing farm advisors, that too much lime induces all sorts of nutrient deficiencies. However, considering my own positive experiences with COF and the writings of Victor Tiedjens, it seems impossible to create a damaging calcium excess with agricultural lime. No matter how much ag lime is spread, calcium saturation will not exceed 85%; at that saturation, food crops still grow excellently because there remains another 15% on the exchange sites to provide plants with more than enough of the other cations. As long as the pH remains below about 7.6, calcium-saturated soil delivers up cations with greater ease than it can when the soil remains acidic; this compensates a great deal for the diminishment of availability due to high pH. I touched upon this topic in Chapter 4 when explaining why repeated light applications of ag lime make my Complete Organic Fertilizer recipe work so well.

William Albrecht's base cation saturation targets — 68% calcium and 12% magnesium — may be highly effective for acidic soils, but they are not the only useful possibility. Another ideal soil model was developed by Victor Tiedjens. It is discussed at length in two of his books, *More Food From Soil Science: The Natural Chemistry of Lime in Agriculture* (1965), and *Olena Farm, U.S.A.: An Agricultural Success Story* (1969). Both titles are available for free download at soilanalyst.org. I urge every food grower to read at least one of them. Tiedjens's ideal soil target is 85% calcium saturation, 5% magnesium, 2%–4% potassium, 1%–2% sodium and 3%–4% "other bases." If those saturation levels are achieved on normally acidic soil by applications of coarsely ground ag lime in the range of 6 to 15 tons per acre, then the soil pH will exceed 7.0 by only a bit. There will be no significant reduction in trace element availability, as happens when soil pH exceeds 7.6. When there are a great many tons of as-yet-undissolved lime present — not merely a 5- or 10-ton surplus, but from 50 to several hundred tons per acre — you have a calcareous soil. With a naturally calcareous soil, the pH will test as high as 8.2 in the lab, but the pH would test lower if that test could be run "live," in the garden and not done with a dried, prepared soil sample. (More on this soon.) If there are also significant excesses of potassium and/or magnesium, the pH can reach 8.5 or 8.6.

Encountering garden soils with a great deal of free lime in them is not unusual. Many backyard growers spread lime every spring and never bother with a soil test. Users of COF have repeatedly applied calcium at a rate of 500–800 pounds per crop; some have done this since the mid-1980s. Eight hundred pounds of elemental calcium equals one ton of high-purity ag lime. If you repeat liming at one ton per acre half a dozen times, all but the heaviest soils will accumulate considerable free lime. The fact that COF has produced excellent results for tens of thousands of gardeners since the mid-1980s proves the workability of Tiedjens's targets to my satisfaction.

Two drawbacks arise from having large quantities of undissolved ag lime in your soil. One is mental; you may consider the presence of this lime to be an excess that needs to be eliminated, so you worry about it. However, this circumstance is only worrisome if you believe that Albrecht's targets are the only ideal targets. If you aim to play Tiedjens's game instead of Albrecht's, there is no problem at all. The other drawback is more significant: having more than a few milliequivalents of free lime present in the soil (remember — 400 pounds elemental calcium per meq) wrecks the usual soil-testing procedures. This can be a serious matter if you unthinkingly set out to balance an "over"-limed soil to Albrecht's targets because you will be working with distorted results if you get the standard soil test.

The first two worksheets in this book (the *Acid Soil Worksheet* and the *Excess Cation Worksheet*) depend on using a Mehlich 3 (M3) extraction. In the M3 test, a soil sample is soaked in an acid about the same strength as white household vinegar; this "extractant" energetically dissolves free lime (just like in the fizz test I'll be describing in a few pages). The M3 method then incorrectly increases the TCEC by the amount of this dissolved free lime. The Total Cation Exchange Capacity is determined from a simple computation that adds up the number of milliequivalents of all major base cations plus any hydrogen (acidity), and what Logan Labs terms the "other bases" attached to the clay-humus. (The next major section, "Calcareous Soils," provides the simple arithmetic used to make this calculation; a look at it should illustrate the previous sentence.)

Suppose a long-established sandy-loam pasture that has not been fertilized or limed in decades is about to be converted to a vegetable garden. It is located in a region where soils normally are acidic. An M3 soil audit shows the TCEC is 10.0, and the soil has 1,920 pounds of calcium per acre on the exchange sites. If it were at our calcium target of 68% saturation, it would hold 2,720 pounds in the furrowslice acre, so we're short by 800 by pounds. To keep this example simple, let's assume the soil already has the ideal quantity of magnesium, sodium and potassium for a TCEC of 10.0.

Instead of spreading the 800 pounds of calcium needed to bring saturation to 68%, the gardener, a follower of Victor Tiedjens, adds four tons of high-purity, #10 ag lime (40% calcium) containing 3,200 pounds of calcium. About one-third of this #10 lime dissolves in the first year, lifting the calcium saturation well over 68% and bringing the pH up close to 7.0; there remains another 2,100 pounds of as-yet-unreleased free calcium in the furrowslice — a bit more than five meq of calcium.

The M3 audit a year later will incorrectly report a TCEC of 15–16, but it is really still about 10.0. The audit will state that the amount of calcium on the exchange sites exceeds 5,000 pounds, but the amounts of potassium, magnesium and sodium will still be about the same as before the lime was spread. But now, weighed against all that calcium, those levels appear to be deficient — if you believe the TCEC actually is 15. If gardeners don't appreciate how a Mehlich 3 extraction operates, they'll robotically apply Albrechtian target saturations for an incorrect TCEC of 15. If they proceed to feed more magnesium and potassium, they'll throw the soil more out of balance — and waste their money as well. If you assume that the TCEC is 15.0 when it really is 10.0, then whatever you add will be in excess of requirements by half-again too much. When you look at the quantities of metals present, particularly copper and zinc, these may seem short against a TCEC of 15, but gauged against an exchange capacity of only 10, they are probably plentiful.

If four tons of ag lime had not been spread, if instead the Albrechtian target of 2,000 pounds of #10 ag lime containing 800 pounds of actual calcium had been added, then next year's M3 test would indicate (still incorrectly) that all or almost all of that ag lime had dissolved and become attached to the exchange points. In reality,

Spotting Signs of Free Calcium

When a Mehlich 3 soil audit encounters more than a small amount of free lime these distortions occur:

- Calcium saturation exceeds 68% while magnesium and/or potassium seem deficient, when in truth, magnesium, potassium, or both may be in excess;
- Soil pH will be 7.0 or higher;
- TCEC may seem too high for the soil type. This effect is especially obvious on otherwise light soils. When you see the first two items on this list on a sandy soil and the TCEC exceeds 10.0, you are probably looking at a free-lime situation.

only about one-third of it had released so far. Because there are a few hundred pounds of as-yet-unreleased calcium present (which would be dissolved by the extractant, and therefore counted as available), the TCEC now adds up to around 11.5 instead of 10, but this is a small error; it is not significant. The way M3 extractions overstate calcium saturation in the presence of free lime can be a useful thing as well as a problem; in this example it would prevent over-liming — the as-yet-unreleased calcium would be indicated as present; the gardener would not be moved to add more.

One good thing about the Mehlich 3 method is that even in the presence of a huge quantity of surplus lime, it still accurately reports the other major cations, as well as the metals and the anions.

What To Do

Having free lime is not necessarily a problem; but it does prompt you to make a choice. You can either: 1) Take steps to eliminate the surplus, restore the soil pH to 6.4, and return to Albrechtian target levels by applying the *Excess Cations Worksheet* to this situation; or, 2) Choose to use Tiedjens's system and get another kind of soil test, one capable of ignoring free lime. At this time, I cannot say with any certainty which is the better choice. My garden is currently divided into more or less equal halves, one half fertilized with the COF recipe in Chapter 4, the other half targeting Albrecht's numbers. Maybe in five years or so, I'll be able to say with confidence which approach grows better food with less difficulty.

If your choice is Albrecht's targets, then in order to eliminate a calcium excess, first get the type of soil test that can accurately determine base saturations in the presence of free lime, then use the *Excess Cations Worksheet,* which will tell you to spread 110 pounds of agricultural sulfur per acre per year. Slowly, the situation will rectify. If, instead, you choose to welcome this calcium, and in fact, to increase it so that calcium saturation reaches the maximum possible level that can be achieved outside of a laboratory test tube — 85% — there are two sub-routes: 1) Henceforth, use the version of Complete Organic Fertilizer that contains ag lime; or, 2) Get the type of soil test that can accurately determine base saturations in the presence of free lime, adopt saturation targets suited to this situation, and proceed to balance the soil. In my opinion, the second option is superior to the first. If you choose to go that way, you're going to create a synthetic calcareous soil. So, you'll want to use the *Calcareous Soil Worksheet.*

Calcareous Soils

If there's enough free lime present to properly label a soil "calcareous," the raw, unamended soil will have a pH above 7.6. Many calcareous soils test pH 8.2 in a lab. At this pH, phosphorus and trace elements are almost completely unavailable. These elements may be present in abundance and could be discovered — if an extraction were done with a much stronger acid than the M3 uses — but at a high pH, they're not accessible to plants and barely appear on the soil audit. Droughty calcareous soils are inevitably low in organic matter. However, acidity develops where there is decomposing organic matter, making phosphorus and other minor nutrients available in those zones. Thus, calcareous soils respond even more strongly to compost than acidic soils do.

A laboratory test of a calcareous garden soil may overstate pH by as much as half a point; the test might indicate 8.2 even though the real in-the-soil pH is actually 7.5 or 7.6. The physics behind this distortion are complex; the bottom line is that the percentage of atmospheric carbon dioxide gas has a huge effect on soil pH when there is a lot of free lime present. At normal atmospheric CO_2 concentration, highly calcareous soils will settle at around pH 8.2. (It'll be higher than 8.2 if the soil carries excesses of magnesium and/or potassium as well as a lot of free lime; I've seen calcareous soils like that test at pH 8.6 — and even 8.7). But, double the concentration of CO_2 in the soil air, and the pH of the same soil drops from 8.2 to 7.6. This phenomena is highly important because lowering pH that much greatly increases the availability of plant nutrients.

When soil contains a significant level of organic matter, it becomes a living entity. Everything living in humus-rich soil is breathing in oxygen and exhaling CO_2, but the soil air does not turn over rapidly, so CO_2 levels increase markedly; it easily reaches levels that are double that of the atmosphere. The actual effective pH of this hypothetical humusy calcareous soil could be around 7.5. But take some of that soil to a testing lab, where they will bring it to scientific dryness (by baking it at around 240°F for a hour or so) and then test the pH; in those conditions, the CO_2 concentration in the soil sample will be whatever it was inside the lab itself — normal atmospheric concentration. So a sample that might test at pH 7.6 if you could test it in position in the garden will test at 8.2 after drying and preparing it for lab work.

What I'm getting at here is a suggestion to generously mix organic matter into calcareous soils. I'd start out a new garden by first digging in a 1-inch-thick layer of compost (or a two-inch thick layer of semi-rotted manure). It would be wise to do this at least one month prior to sowing the first seeds. Keep the soil moist during that time, giving the soil a chance to develop a much-invigorated biology. And you want to give the ground a chance to settle into a new pattern before doing any soil testing.

Calcareous soils often present severe phosphorus deficiencies. Phosphate may be present, but it's hidden because of high pH and all that calcium to combine with. For that reason, mixing OMRI-approved insoluble phosphate fertilizers into calcareous soil is usually wasted effort and money. These fertilizers need actual acidity to release phosphate. Better to brew phosphates into active, moist compost for a month to "complex" the phosphate with organic matter. Then concentrate this phosphorus-fortified compost immediately beneath or next to the plants or rows. This practice, called "banding," creates a zone of excess P within a zone of lower pH — where there is a greater likelihood of plants finding available nutrition of all sorts. Banding is also a frugal practice.

Here's how to band: When planting in rows, make a broad furrow that is four to six inches deep; do this with a large hoe. Fortified compost is placed in the bottom of that "V," and the small trench is refilled with the soil originally drawn out by the hoe. Then, a shallow furrow is made immediately adjacent to or above the subterranean band; there, seeds can safely germinate. As the seedlings start developing, their roots encounter superfertility.

When setting out seedlings for what will become large plants (tomatoes, Cucurbits or big cabbages, for example), first dig a small

Foliar Feeding

Farmers foliar feed in one of two ways. A fast pass with a tractor spray rig (or crop duster) covers a field with a certain amount of material dissolved in sufficient water to dampen most of the leaves over a predetermined area. In a fast pass, every plant gets about the same amount of material, but every leaf may not. The concentration of the solution when doing this sort of spraying has to be low enough that droplets will not burn leaves but high enough that a partial covering makes a sufficient dose. Generally, instructions will say to mix so many pounds of material in 20 gallons of water, and then cover one acre with it.

The garden method is usually done with a hand-pump pressure sprayer. I use either a five-quart model that I carry in one hand, or else a backpack rig that holds about three gallons. In desperate straits, you could use a whisk broom or large paint brush repeatedly dipped into a bucket of solution and then rapidly swung so as to spray droplets on the leaves. Foliars should be applied when the leaf stomata (breathing pores) are open, which only happens when the sun is not strong — and it happens best when there is weak sun and high humidity. The very best time of day for foliar spraying is usually early morning. Next best is when the day is heavily overcast or an hour before dark, if it has cooled down enough. It takes me about one quart of water to cover 100 square feet of bed when the crop there has achieved a dense leaf canopy. You can test that assertion out for yourself by seeing how much area you can cover while spraying one quart of plain water.

Some species, especially the Brassicas, have waxy leaves that repel moisture. Foliars bead up and run off Brassica leaves without penetrating. Sophisticated farmers use chemical spreader-stickers that increase nutrient uptake. Essentially, these are surfactants. You can get nearly as good a result by adding one-quarter teaspoonful of dishwashing detergent per quart of spray.

For iron, zinc and manganese sulfates, the usual foliar application is half a pound of elemental per acre, per spray. For copper, use half of that rate — one-quarter pound of elemental copper per acre per spray. At that concentration, spraying once a month should be enough, but you can do it more often if the plants stop showing a benefit before a month passes.

I work out the proper spray concentration this way: One acre is 43,560 square feet. That means there are 435 areas of 100 square feet each in one acre; actually, ☛

it's easiest to reckon there are either 400 areas (if you want to be a bit generous) or 450 (to be a bit scant). The worksheets conveniently list the percentage compositions of the sulfates. For example, zinc sulfate is 35% elemental zinc. So, to spread one-half pound of elemental zinc per acre in the form of zinc sulfate, you need to uniformly distribute 1½ pounds of zinc sulfate over 400 or 450 areas of 100 square feet each.

I find it far easier to work out dosages and rates of application in grams, rather than in fractions of an ounce or quarter-teaspoons. There are about 450 grams per pound, so there are 675 grams in 1½ pounds. Evenly apportioning 675 grams of zinc sulfate over 450 beds means applying 1.5 grams per bed. So, if I dissolve 13.5 grams of zinc sulfate (anywhere from 12–15 grams will serve) into nine quarts of water and spray that water more or less evenly on nine beds of 100 square feet each, making sure that water drips off all the leaves, then I will have given each of those nine beds a dose of one and one-half grams of zinc sulfate, which works out to be roughly the rate of one-half pound elemental per acre. Don't stress: there's no need for perfect accuracy here. Half again more would do no harm, although half less might not be effective. You could just about as well put in a tablespoon of any of the elements in sulfate form (except copper, which can be caustic; it needs greater dilution and a smaller per-acre application rate) into a gallon of water and spray it heavily over as much area as it covers.

Mike Kraidy suggests that the best way to foliar feed on a home-garden scale is to do it with a watering can, not only wetting all the leaves, but simultaneously drenching the root zone. The concentration is the same when you soil drench, but going at it this way, you'll need several times more solution.

hole about six inches deep and a foot around by removing a few shovel's full of soil from a spot. Then put a large shovel's worth of fortified compost (and other fertilizers if you wish) into that hole, dig these materials in a bit so they dilute into the soil but are still concentrated, and then set the seedling right on top of that spot.

If you practice banding, avoid those superfertile zones when you next take soil samples. Furthermore, I suggest that after the crops have finished and before you take new samples, dig up the entire area to distribute the concentrated nutrients.

Do not band lime, gypsum, potassium, magnesium or agricultural sulfur; these work better when mixed throughout the soil.

Without Doing a Soil Test

If you don't want to do a soil test, I suggest that you handle calcareous soils this way:

- Broadcast the usual quantity of COF. If you live in a semi-arid or arid region and have naturally "fizzy" soil (i.e., soil that has a lot of calcium already), use the COF recipe for low-rainfall areas, because there is no need for any more ag lime in your soil. If you are making a synthetic calcareous soil out of a normally acidic one in order to practice the Tiedjens method, then use the recipe for humid districts — the one containing equal parts of ag lime and gypsum. These recipes are found in Chapter 4.
- Cover the area with compost one inch thick or rotted manure two inches thick.
- Dig it all in.
- During the growing season, repeatedly foliar feed trace elements to the entire garden. Start doing it once a week, keep an eye on your results, and hope you can do it less often than that. You can create your own foliar sprays containing zinc, copper, manganese and iron sulfates, and you can spray liquid kelp to provide micronutrients. Otherwise, foliar feed the most complete and balanced liquid fertilizers you can find (and that satisfy any organic-certification requirements you might have). If you have any doubts about the importance of foliar feeding when growing on calcareous soil, please look again at Figure 8.1 to see how nutrient availability decreases as pH increases.
- If you live in an arid or semi-arid region, you will need to leach the soil once a year because your irrigation water almost certainly contains high levels of alkaline salts (sodium, magnesium and calcium). This is best done immediately after you get some meaningful rain. Thoroughly rinse out excess salts by spreading enough water to soak in at least 18" deep — twice that depth is even better. If your soil has a lot of clay in it, it could take more than a foot of water applied at

one time to effectively leach the garden.

- In subsequent years, reduce organic matter to ½-inch-thick additions.

The Fizz Test

The OSU Extension Service advises people living east of the Cascades (a region where calcareous soils are the usual thing) to confirm they have a calcareous soil by doing a vinegar fizz test. The degree of fizzing reveals approximately how much free calcium is present. Anyone gardening in semi-arid or arid country or in a region known to have limestone-derived soils should routinely do a fizz test before deciding which sort of laboratory soil test to get.

Put a few drops of household white vinegar on a spoonful of dry agricultural lime; it will fizz. Put an ounce of dry soil into a bowl. Add a tablespoon of white vinegar. If there is an abundance of free calcium there, the vinegar will make it fizz. How strongly it fizzes roughly indicates the amount of calcium present. If there's a lot of free calcium, it'll fizz noticeably. If there is only a small amount of free calcium, you may have to put your ear next to the soil to faintly hear the bubbling. If you hear or see any fizzing, then the M3 soil test is not all you need. Nor is the usual ammonium acetate test your answer, though that is the test most conventional farm consultants would advise you to get. I'll explain shortly.

Standard agronomy says the pH of calcareous soil should be lowered by heavily amending the land with agricultural sulfur. You fizz-test the soil each spring, and if it still fizzes, spread and till in another 50

Fizz test result	Estimated Carbonates Present (%)	Annual addition of elemental sulfur (ton/acre)	Duration (years)
None	0	none	none
Heard (barely audible)	0-1	0.5-1	1
Slight (few bubbles)	1-2	1	1-2
Moderate (several bubbles)	2-3	1	2-3
Vigorous (many bubbles)	>3	1	3+

Table 8.1.

pounds ag sulfur per 1,000 square feet (2,000 lb/acre, which amounts to 20 times the level I ever suggest as a maximum dose), and then repeat the fizz test again next year. Eventually, the vinegar no longer fizzes, and your topsoil is at the point where its cations can be brought into an Albrechtian balance according to M3 soil test results. See Table 8.1 to see what the OSU Extension Service says about how many years it will take you to get the pH in order.

Raw sulfur in moist soil is converted by soil bacteria into sulfuric acid that reacts immediately with free calcium, forming gypsum. Gypsum is a lot more soluble than ag lime is. This conversion happens effectively only when the soil is warm. In eastern Oregon, this is a significant consideration because the soils only warm up enough to convert sulfur during the peak summer months. Gypsum and the other salts formed can be leached out once or twice a year. If the soil has such poor drainage that it can't be leached, you won't be able to eliminate the free calcium.

I consider the entire approach of lowering pH with sulfur to be far from desirable. For one, it can take several years for an application of sulfur to be fully converted to sulfate. The annual fizz test may prompt you to add even more sulfur before the one or even two year's worth of applications are done reacting. I interviewed the extension agent who wrote the OSU Bulletin on this. He admitted that if you add sulfur year after year, it is not unusual to overshoot the mark — and get soil with a pH as low as 5.0. Then you have to add lime to bring it back up. This whole approach seems unnecessarily harsh, unless your goal is to grow rhododendrons or blueberries. It is also expensive. One ton of ag sulfur costs around US$900. I wouldn't do it; I suggest you don't, either. Besides, a pH of 8 still grows good fruit and vegetables.

Balancing Calcareous Soils

I thought long and hard about including this section in this first edition. I do not feel completely ready to write it. I have had only six soil samples tested in the manner I am about to describe — they were from Kentucky (from the "Blue Grass" region), eastern Texas (chalk) and upstate New York (this soil was intentionally "over"-limed by a

Tiedjens follower). But the deadline is hard upon me. If I do not share what I know now, I may not have another chance. Even if this book goes into another edition, it may be two years before that happens. So best I point you in what I think is the right direction.

There's something intuitively satisfying about analyzing a Mehlich 3 soil audit because the numbers it reports closely match what is actually in the soil. When the test says you have 23 pounds of available zinc per furrowslice acre, you really do have close to that. And if your target calls for 50 pounds of available zinc, then by adding 27 more pounds, you feel you have done the right thing — and it works, too. However, the Mehlich 3 soil audit uses an acidic extractant that dissolves a great deal of free calcium during the extraction process (just like the fizz test does). But the M3 reports this dissolved free calcium as being available — as calcium attached to the TCEC — when it really isn't. Then, by including that calcium, the Mehlich 3 arithmetic generates a falsely high TCEC — sometimes an enormously elevated one. A professional soil analyst/farm consultant who routinely works with calcareous soils would never try to balance a soil using a Mehlich 3 extraction. There is an appropriate test available, but so far only a few professional soil analysts are familiar with it. I expect this situation will improve as this new knowledge about soil and nutrient-density is circulated more widely.

The test to use on calcareous soils uses an elevated pH ammonium acetate extraction. This test can accurately determine the four base cations in calcareous soil. However, the ammonium acetate (AA) test does a poor job on the anions and metals in a calcareous soil; the M3 does accurately report these. By combining the strengths of these two tests, it is possible to balance calcareous soil. So, it is a smart idea to do a fizz test before you send off a soil sample. If the soil fizzes, you can't balance it with M3 results. If you already ordered an M3 test, you'll need an additional test and will have wasted the cost of the first attempt. So you'll save a few dollars by ordering both at once and speed up the process too. Don't worry: it doesn't involve a lot of money one way or the other. As I write this, Spectrum Analytic in Ohio charges another four dollars for the elevated pH ammonium acetate extraction when done at the same time the M3 is run.

The Correct Extractant

At pH 8.2 or higher, free calcium carbonate cannot feed the plants no matter how much is present because at or above pH 8.2 it is not possible to make calcium carbonate dissolve. At that pH, free calcium becomes part of the inactive soil nutrient reserve of a calcareous soil, similar to undissolved grains of other sorts of rock minerals. Calcium carbonate normally dissolves very slowly in mild soil acids. Although it is far more readily soluble than most other rock fragments, it is not soluble at pH 8.2; it cannot add any effective amount of calcium to the soil solution as long as the pH is so high. And that's why, ironic as it seems, the most harmless and entirely useless thing you can amend calcareous soil with is agricultural limestone. It does nothing. It's like adding fine sand to an already sandy soil.

The trick to getting accurate results on calcareous soils is to not dissolve any free calcium while doing the extraction. The majority of North American soil labs I've encountered so far do ammonium acetate extractions with the extractant brought to a neutral pH of 7.0. This method forms the backbone of modern industrial agriculture; it provides conventional farm advisors with useful numbers for creating big yields. The AA at pH 7.0 is much like the M3; it reads available calcium accurately only when there's little or no free calcium present. When there is a lot of free lime, the usual AA extraction, being a neutral pH extractant instead of an acidic one, overstates calcium somewhat less than an M3 would. On a sample where the M3 audit would report 10,000 pounds of calcium per acre, an AA extraction at pH 7.0 might report 7,000 pounds. But the true quantity might be 4,000 pounds.

A lab routinely using AA extractions at pH 7.0 gives recommendations intended to produce maximum yield with no "wasted" input. I do not use AA labs because nutrient-density peaks only when all element levels (except potassium) are very much higher than what it takes to produce peak yield — and they must be in balance. An ammonium acetate test does not reveal amounts of the anions or metals with an accuracy even close to what an M3 does. To balance soil, we want to know how many pounds of each element are actually present in the furrowslice acre. The only way I know to get this information is through an Mehlich 3 extraction.

However, an elevated pH AA extraction at pH 8.2 does not dissolve any free lime. It gives accurate numbers for all four major cations in fizzy soil, but is rather useless for balancing the metals and anions. Undaunted by this minor complication, we'll simply use the most useful aspects of each of these extraction methods. If you'd like to know more about this subject, look up the paper published by the United States Golf Association by Brian Whitlark, "Soil Testing Procedures for Calcareous Soils," *USGA Green Section Review* Vol. 49 (9), March 4, 2011. It's available on the Michigan State University Library's website at: http://gsr.lib.msu.edu/article/whitlark-soil-3-4-11.pdf.

The Correct Soil Lab

I have not assessed more than a handful of the hundreds of soil labs in North America. At present, I only know two labs suitable for balancing calcareous soils, Spectrum Analytic and Logan Labs. With Spectrum, request an "S3 test plus an additional elevated pH ammonium acetate extraction of the four major cations." With Logan Labs, you want their "standard M3 soil test with the four bases extracted with ammonium acetate at pH 8.2."

Here's an example of how it works.

Dave lives in upstate New York; he owns a hobby farm with mostly sandy-loam soils. He started out a few years back as a follower of Victor Tiedjens and then confused himself by doctor-hopping to Michael Astera's ideal soil system. He first spread ag lime at many tons per acre. Before liming, he got a Logan Labs test that said he had an acidic soil with an exchange capacity around 10.0; if he added a great deal of organic matter over many years, the TCEC could be expected to increase to around 15. This spring, when Dave sent samples for M3 audits, the reports came back looking as shown in Figure 8.5.

Take a look at the column labeled "B." The numbers are typical of what an M3 audit looks like when there is a lot of free calcium present. Soil pH is elevated (7.3); there seems to be large deficiencies in potassium and magnesium — because they're being gauged against a false TCEC of 34.79 — and calcium saturation is incorrectly reported at 88% of that TCEC. Dave was confused, and more than just a bit upset.

How was it possible that after putting in so much fertilizer over the preceding few years he came up so short in K and Mg; and how did the TCEC increase three-fold?

Soil Report

Job Name ______ Date 3/30/2012

Company ______ Submitted By ______

Sample Location			B					
Sample ID								
Lab Number			168					
Sample Depth in inches			6					
Total Exchange Capacity (M. E.)			34.79					
pH of Soil Sample			7.30					
Organic Matter, Percent			4.21					
ANIONS	**SULFUR:**	p.p.m.	26					
	Mehlich III Phosphorous:	as (P_2O_5) lbs / acre	3638					
EXCHANGEABLE CATIONS	**CALCIUM:** lbs / acre	Desired Value	9462					
		Value Found	12296					
		Deficit						
	MAGNESIUM: lbs / acre	Desired Value	1001					
		Value Found	545					
		Deficit	-456					
	POTASSIUM: lbs / acre	Desired Value	1085					
		Value Found	184					
		Deficit	-901					
	SODIUM:	lbs / acre	52					
BASE SATURATION %	Calcium (60 to 70%)		88.37					
	Magnesium (10 to 20%)		6.53					
	Potassium (2 to 5%)		0.68					
	Sodium (.5 to 3%)		0.33					
	Other Bases (Variable)		4.10					
	Exchangable Hydrogen (10 to 15%)		0.00					
TRACE ELEMENTS	Boron (p.p.m.)		0.83					
	Iron (p.p.m.)		129					
	Manganese (p.p.m.)		29					
	Copper (p.p.m.)		1.73					
	Zinc (p.p.m.)		74.32					
	Aluminum (p.p.m.)		604					
OTHER								

Fig. 8.5. **_Logan Labs, LLC_**

I asked Dave to send another sample to Spectrum for their usual M3 audit plus an AA extraction of the bases done at pH 8.2. The report that came back is shown in Figure 8.6.

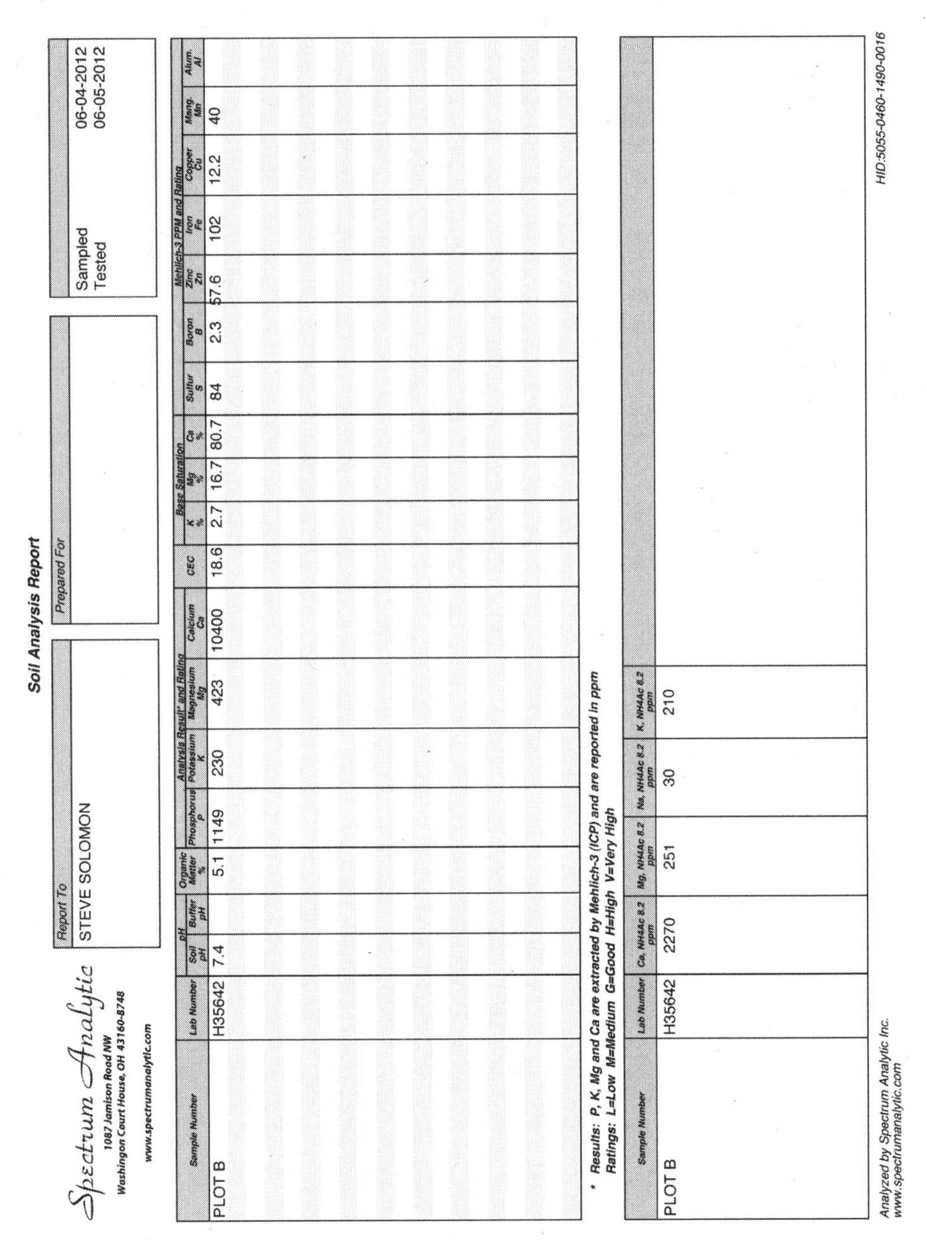

Spectrum Analytic
1087 Jamison Road NW
Washington Court House, OH 43160-8748
www.spectrumanalytic.com

Soil Analysis Report

Report To: STEVE SOLOMON

Prepared For:

Sampled 06-04-2012
Tested 06-05-2012

Sample Number	Lab Number	pH: Soil pH	pH: Buffer pH	Organic Matter %	Analysis Result* and Rating: Phosphorus P	Potassium K	Magnesium Mg	Calcium Ca	CEC	Base Saturation: K %	Mg %	Ca %	Mehlich-3 PPM and Rating: Sulfur S	Boron B	Zinc Zn	Iron Fe	Copper Cu	Mang. Mn	Alum. Al
PLOT B	H35642	7.4		5.1	1149	230	423	10400	18.6	2.7	16.7	80.7	84	2.3	57.6	102	12.2	40	

* Results: P, K, Mg and Ca are extracted by Mehlich-3 (ICP) and are reported in ppm
Ratings: L=Low M=Medium G=Good H=High V=Very High

Sample Number	Lab Number	Ca, NH4Ac 8.2 ppm	Mg, NH4Ac 8.2 ppm	Na, NH4Ac 8.2 ppm	K, NH4Ac 8.2 ppm
PLOT B	H35642	2270	251	30	210

Analyzed by Spectrum Analytic Inc.
www.spectrumanalytic.com

HID:5055-0460-1490-0016

Fig. 8.6.

Spectrum uses a proprietary method to somewhat reduce overstatement of calcium levels when doing M3 audits; it also adjusts down TCEC on heavy soils, so in this case, Spectrum reported an M3 TCEC of only 18.6, a figure that's a lot closer to reality than what Logan reported. Still, we aren't going to use Spectrum's M3 results for the four bases. In the section at the bottom of the report form, you can see the levels discovered by the ammonium acetate extraction.

So what does Dave do with this information?

The *Calcareous Soil Worksheet*

There is a third worksheet especially for calcareous soils and for intentionally "over"-limed soils (you'll find it in the Appendix). Dave's first step is to fill in the *Calcareous Soil Worksheet,* shown in Figure 8.7. Notice that Spectrum does not report levels in pounds per acre (as Logan does, for the convenience of amateurs), but instead, reports levels in parts per million, a method much preferred by professional analysts. Recall please that 1 ppm = 2 lb in a furrowslice acre. In the example illustration, I used M3 levels for all anions and metals; I used the ammonium acetate extraction (bottom row) numbers only for the four major cations.

To calculate TCEC the first step is to convert ppm to lb/ac and then use the following formula (it also appears at the bottom of the *Calcareous Soil Worksheet.*

Calculating TCEC:

$$\frac{\frac{\text{lb/ac calcium}}{400} + \frac{\text{lb/ac Mg}}{240} + \frac{\text{lb/ac K}}{780} + \frac{\text{lb/ac Na}}{460}}{(100 - \text{percent H+} - \text{other bases})} \times 100 = \text{TCEC}$$

In the case of calcareous soil, there is no H+ and other bases usually are about 4%.

Here's the calculation for Dave's Plot B:

- Ca: 2270 x 2 / 400 = 11.35
- Mg: 251 x 2 / 240 = 2.09
- K: 210 x 2 / 780 = 0.53

- Na: 30 x 2 / 460 = insignificant
- Subtotal: milliequivalents of four base cations is 13.97.

Calcareous Soil Worksheet
U.S. Measurements

Name *Dave R.*
Plot or Field *Plot B*
Date of Test *06-05-2012*

Sample Depth 6 inches All numbers on this worksheet assume a six inch sample depth

pH *7.4* This worksheet is for soils that naturally hold free calcium (usually pH over 7.5) and those artifically created "Tiedjens" style (usually pH 7.1 or 7.2). Check that you actually have calcareous soil by doing a Fizz Test. Then get the proper soil test.

TCEC *14.5* Use the results from an elevated pH ammonium acetate extraction to determine TCEC. If necessary, calculate TCEC yourself using the formula on the bottom of this worksheet. Do not use levels discovered by a Mehlich 3 extraction or an ammonium acetate at pH 7.0 extraction to determine TCEC on calcareous or"over" limed soils.

Organic Matter % *5.1* From normal soil organic matter decomposition, assume approximate release of N = 15-25 lb N per 1% OM. Varies with temperature, moisture and soil air supply. N = 0.22 x NO_3

Element	Actual Level		Calculating Target Level Pounds per acre	Target Pounds per acre	Deficit Pounds per acre
Sulfur **S**	ppm	*84*	**S minimum = Mg (Target Level)**	*174*	*6*
	lb/ac	*168*			
Phosphorus **P**	ppm	*1149*	**P = K (Target Level)** Calculate using actual P, not phosphate. P = 0.44 x P_2O_5	*300*	*Excess*
	lb/ac	*2298*			
Calcium **Ca**	ppm	*2270*	**TCEC x 400 x 0.85 = Target Level**	*4930*	*390*
	lb/ac	*4540*			
Magnesium **Mg**	ppm	*251*	**TCEC x 240 x 0.05 = Target Level**	*174*	*Excess*
	lb/ac	*502*			
Potassium **K**	ppm	*210*	**K is proportional to TCEC: see chart**	*300*	*Excess*
	lb/ac	*420*			
Sodium **Na**	ppm	*30*	**TCEC x 460 x 0.01 = Target Level** Be certain of good water quality before adding sodium	*67*	*Ok*
	lb/ac	*60*			
Boron **B**	ppm	*2.3*	**B = 2 lb/ac if CEC below 10** **= 4 lb/ac if CEC above 10**	Do not exceed 4 pounds *4*	*Excess*
	lb/ac	*4.6*			
Iron **Fe**	ppm	*102*	**Fe = 100 lb/ac if CEC below 10** **= 150 lb/ac if CEC above 10**	*150*	*Ok*
	lb/ac	*204*			
Manganese **Mn**	ppm	*40*	**Mn = 55 lb/ac if CEC below 10** **= 100 lb/ac if CEC above 10**	*100*	*20*
	lb/ac	*80*			
Copper **Cu**	ppm	*12.2*	**Cu = ½ Zn (Target Level)**	*15*	*Excess*
	lb/ac	*24.4*			
Zinc **Zn**	ppm	*57.6*	**Zn = 1/10 P (Target Level)**	*30*	*Excess*
	lb/ac	*115.2*			

Potassium Target Levels

TCEC	Pounds	TCEC	Pounds	TCEC	Pounds
		16	308	28	394
		17	316	29	397
		18	324	30	400
7	201	19	332	31	403
8	212	20	340	32	406
9	225	21	348	33	409
10	240	22	356	34	412
11	252	23	364	35	415
12	264	24	372	36	418
13	276	25	380	37	420
14	288	26	384	38	422
15	300	27	388	39	424

If TCEC is lower than 7, use value for 7. If it is over 39, use value for 39.

	One acre, six inches deep weighs	One hectare, 80 mm deep weighs
1 meq Calcium	**400 lb**	**400 kg**
1 meq Magnesium	**240 lb**	**240 kg**
1 meq Potassium	**780 lb**	**780 kg**
1 meq Sodium	**460 lb**	**460 kg**

1 ppm = 1mg/kg = 2 pounds/acre = 2.24 kg/hectare

Calculating TCEC:

$$\frac{\frac{\text{lb/ac calcium}}{400} + \frac{\text{lb/ac Mg}}{240} + \frac{\text{lb/ac K}}{780} + \frac{\text{lb/ac Na}}{460}}{(100 - \text{percent H}^+ - \text{other bases})} \times 100 = \text{TCEC}$$

In the case of calcareous soil, there is no H^+ and other bases usually are about 4%.

Fig. 8.7.

There is no H+ when pH is over 7.0.

Other bases: Spectrum doesn't give this figure, but it is inevitably around 4%.

Therefore, this Dave would calculate his TCEC as: 13.97/96 x 100 = 14.55.

From this point on, the *Calcareous Soil Worksheet* is completed much like the others. Note that target levels for anions and metals are somewhat lower than for acidic soils. Also, note that application limits on this worksheet are more restrictive because the TCEC already is largely saturated with calcium. The main point is still to bring the four major cations into balance, but the targets are different.

So, to figure out his targets, Dave would just fill the numbers from the test result into the formulas on the worksheet. Thus, Dave's target calcium is: 14.5 (the actual TCEC) x 400 x 0.85 = 4,930 pounds. What do you know! Despite all that pre-existing lime, calcium saturation is still 390 pounds short of the target amount.

Dave has 502 pounds of Mg on a TCEC of 14.5. To calculate the magnesium target, it's 14.5 x 240 x 0.05 = 174. What do you know! We have a large magnesium surplus when gauged against a target of 5% saturation. Target sodium is worked out as usual, although at 1% saturation, not 2%, as on acidic soils. Target potassium comes from the chart (rounding the TCEC up, to 15): 300 pounds. When this amount is put into the worksheet, we discover there's a significant excess of potassium. And clearly, it is because of those two excess bases that the soil pH is 7.4 instead of 7.1. From this point, working out the targets for everything else proceeds as usual.

The Prescription

Dave has three deficits: 6 lbs of sulfur, 390 lbs of calcium and 20 lbs of manganese. We turn to side two of the *Calcareous Soil Worksheet* to figure out how to provide for those deficits (Figure 8.8).

Dave needs to provide 390 pounds of calcium. To get that he'll spread gypsum at 1,902 pounds per acre (gypsum contains 20.5% elemental calcium).

There is no sulfur application limit on this worksheet, but there is a minimum level. Sulfur should at least equal the weight of target magnesium (at 5% base saturation). The best way to supply that requirement is first from any sulfate fertilizers required, and then with gypsum

	Deficit From other side of worksheet	Application Limit Per acre/year	Quantity and Material to add	S	Mg	Ca
Sulfur **S**	*6*		If no other sulphates needed, use gypsum to reach minimum target level.			
Phosphorus **P**	*Excess*	175 lb/ac elemental P	Use Soft Rock Phosphate if you are not willing to use Monoammonium phosphate.			
Calcium **Ca**	*390*	Gypsum: 1 ton per acre	Use gypsum; it is okay to exceed minimum sulfur target. *Gypsum 1902 lb/ac*			
Magnesium **Mg**	*Excess*	No more than 20% of target magnesium per year	Use K-Mag or Langbeinite even if this puts K or S into excess.			
Potassium **K**	*Excess*	100 lb/ac elemental K	Use potassium sulphate. If this puts sulfur over the target level, go ahead anyway.			
Sodium **Na**	*ok*					
Boron **B**	*Excess*	2 lb/ac elemental B				
Iron **Fe**	*ok*	Foliar feeding only				
Manganese **Mn**	*20*	No more than 10 lb elemental Mn	*Manganese sulfate 30 lb/ac*			
Copper **Cu**	*Excess*	No more than 5 lb elemental Cu				
Zinc **Zn**	*Excess*	No more than 10 lb elemental Zn				

	N	P	K	S	Ca	Mg
Fish Bone	4	8.8		.06	19.0	.03
Fish Meal	10	2		0.6	2.3	.03
Crab Shell	3	1.5	.025	.02	23.0	1.3
Blood Meal	13	0.5				
Feather Meal	12	0.0	0.35	0.4	0.6	
Bone Meal****	3	13.0		2.5	12	0.3
Oilseed Meal	6	1.5	1.0			
Copra Meal	4	1	0.7			
Kelp Meal	1	0.3	2.5	2	2	0.7
Ag Lime					32-39	2
Dolomite					22	13
Gypsum				17	20.5	
Oyster Shell					36	0.03
Magnesium Oxide						50
Montana Hard Rock Phos**		1.3			29	
Calphos		8.8			20	
Monoammonium Phosphate		23 (Plus 12% N as NH^3)				
K-Mag			18.2	22		11
Langbeinite			15.6	23		12
Greensand***		.05	6	1.3	1.5-3.0	2-4
Ag Sulphur				90		

Sea Salt	35%	Sodium (Na)
Borax	10%	Boron (B)
Iron Sulfate	18% S	30% Fe
Manganese sulfate	19% S	32% Mn
Copper Sulfate	12.5% S	25% Cu
Zinc Sulfate*	17% S	35% Zn
Potassium Sulfate	17% S	42% K
Magnesium Sulfate	13% S	10% Mg

* Zinc sulfate picks up moisture from the air; store in airtight container.

** Hard Rock Phosphate is 1.5% available P and contains around 27% insoluble phosphate.

*** Greensand contains 9% Fe, 50% Si and many trace elements. More than half its potassium content is insoluble.

**** Bonemeal contains 5.7% sodium.

Date of Issue: 07/07/2012

Fig. 8.8.

which contains 17% sulfur. 1,902 pounds of gypsum (which will be used to provide the calcium) also provides 323 pounds of sulfur; there'll be no problem meeting our target minimum.

The soil is short the manganese target by 20 pounds, but there is an application limit of 10 pounds. Dave will spread his 10 pounds elemental Mn in the form of 31 pounds manganese sulfate ($MnSO_4$) per acre. (Figured using the worksheet's list of the percentage compositions of the sulfates.)

The standard prescription also is applied: one hundred pounds per acre of nitrogen and some kelp meal or Azomite for micronutrients. These elements are not on the worksheet, but they should not be overlooked.

And that's Dave's prescription.

Restrictions and Special Conditions

Calcium: No harm comes from adding agricultural lime to calcareous soils, but no benefit accrues either, unless it is an artificially created calcareous soil that has not yet exceeded 80% calcium saturation. When additional calcium gets delivered to calcareous soils in the form of gypsum, there's great benefit. If calcium saturation is below the target of 85% saturation — which can happen on highly calcareous soil when there also is excess magnesium, potassium and/or sodium on the TCEC — then adding gypsum to boost calcium saturation helps leach out excess cations. There is an overall gypsum application limit of one ton per acre. Do not worry if applications of gypsum put sulfur over the target level because the *Calcareous Soil Worksheet* has no sulfur limit or target, only a minimum amount.

The most amazing thing about calcareous soils — the hardest part for most people to accept or grasp — is that even though the soil contains an enormous amount of calcium, the plants themselves can experience a shortage of calcium nutrition. There are two ways to prove this is the case. One is to analyze leaf samples. It is well known what the sap levels of the various nutrients should be, and it is not unusual to have results show serious shortages of calcium nutrition in calcareous soils. The other way to demonstrate a shortage of available calcium is to add gypsum and see if it makes the plants grow better.

That's one reason there's a large dose of gypsum in the version of COF intended for neutral and calcareous soils.

When maximum possible calcium saturation levels are intentionally being built on what would otherwise be acidic soil, in the event calcium saturation tests below 80%, enough ag lime should be added to bring saturation to at least 80% and enough gypsum also added to lift the saturation from 80% to 85% (up to one ton). Normally, if calcium saturation has not reached 85%, there will be surplus magnesium and/or potassium. Gypsum will simultaneously reduce these excesses.

Magnesium: The target saturation is 5%. Mike Kraidy is a farm consultant who is very experienced with calcareous soils. In his experience, the only fertilizers that effectively raise magnesium saturation in calcareous soils are K-Mag or langbeinite. Foliar feeding Epsom salts at one tablespoon per gallon (or, even better, drenching the soil immediately around the plants with that solution while foliar feeding) is a good preventative of magnesium deficiency in soils that test below 5%. K-Mag also contains potassium. If using K-Mag to build magnesium puts potassium into excess, go ahead and use it anyway. The magnesium it releases works to displace surplus potassium.

Potassium: The *Calcareous Soil Worksheet* has its own table of potassium levels, intentionally set at lower targets than on the other two worksheets. Potassium sulfate, K_2SO_4, is the best potassium fertilizer. There is an application limit of 100 lb/ac elemental potassium. Providing adequate potassium is crucially important, especially so because high pH makes potassium go relatively unavailable.

Sodium: In arid and semi-arid climates, irrigation water commonly carries large quantities of sodium. Do not add sodium without first obtaining an analysis of your irrigation water. If your water contains a lot of magnesium, you may find it nearly impossible to completely reduce excesses of this element against the constant additions from your irrigation, but ongoing applications of gypsum will keep magnesium saturation in check. Mike Kraidy told me about how irrigation water in arid climates can be so loaded with salts that several tons of gypsum per acre are required every single year to leach them out. Growers in that situation use a super-fine grind that can be mixed into the irrigation water itself. I hope none of my readers have to cope with water like that.

Anions: Providing bountiful phosphorus is particularly difficult on calcareous soils because when it combines with calcium, it goes unavailable. Mixing organic phosphate fertilizer with high-quality compost makes applied P far more effective. Soft rock phosphate, when it is first assimilated by a brewing compost heap, works excellently. If you must put phosphate fertilizer straight into the soil, monoammonium phosphate is the best choice.

Trace elements (metals): If the soil tests short on manganese, copper and/or zinc, best to use a split approach: put some into the soil (the very best way to do that is banding plants with the metal after it has been mixed with compost). You should also foliar feed additional amounts through the crop cycle. Note that there are stricter soil application limits for calcareous soils than for other soil groups.

Iron: This element rapidly becomes unavailable in calcareous soils. Putting soluble iron into high-pH soil rarely does any good. For that reason, if a soil audit shows iron as deficient, do not add iron sulfate to the soil; plan on foliar feeding it. Organic matter often brings with it large quantities of iron; in fact, building soil organic matter levels can be a big help in providing for trace element deficiencies.

Chapter 9

Compost

In the early 1990s, I wrote *Organic Gardener's Composting* for my friend George Van Patten. He ran a small garden book publishing business that lacked a basic compost-making guide. A year or so after publication of my book, George left the States to live in Spain; his publishing business evaporated, and my book went out of print. Now, I control its distribution, so you can download a free copy at soilandhealth.org.

Organic Gardener's Composting is a complete beginner's guide. If you're a novice who has never carefully investigated making compost, you need to know the stuff that's in there. If getting a free ebook off the Internet doesn't appeal, there must be a dozen general compost-making guides in print right now; your local library probably has some, as well as a few dozen others that are out of print. Most, if not all, will have diagrams for building bins and step-by-step recipes for layer-cake heaps. They will have tables listing carbon-to-nitrogen ratios and the average nitrogen contents of manures. They'll introduce you to worm bins, tumblers and the other usual side-paths. If you've never read a book on composting, you should. In fact, if you haven't, and if you haven't already made a few compost heaps, what I am about to say in this chapter won't be of much use to you. However, what's to come can advance an experienced composter a great deal in relatively few pages.

Basic composting books imply that all compost is good compost. That you can go about making compost in a variety of ways, but the end result of all these approaches will be powerful, effective compost.

This false notion has prevented many organic gardeners from ever achieving nutrient-dense results — or even an effectively productive garden, for that matter. The truth is, much home-garden compost is ineffective. Its carbon-to-nitrogen ratio is too high, usually because the starting C:N was too high, and its mineral content is unbalanced. It usually does not produce a strong growth response. Because of this, gardeners naturally use much more of it, trying for a growth response; in the process, they put their soil further out of balance.

Another major lack: home-garden composting guides do not address how much organic matter soil needs in the first place. This should be a major concern. Instead, gardeners are advised that there is really no limit to how much compost they might want to make and use. These books assert that compost should be the only source of soil improvement, and if the garden don't grow well enough — put some moron-it. You'll find a thorough discussion of this topic in *Organic Gardener's Composting*. Even though that book was written in the early 1990s, my opinions on the topic haven't changed.

The confusion highlights a major blind spot in the organic tradition. Supposedly, large additions of organic matter are required to loosen soil and allow it to hold enough air to grow crops well. This assertion is true — in a way. High-enough levels of organic matter will loosen tight soil. But this fix becomes a tedious treadmill — and worse, it does not lead to nutrient-dense food. It was natural for the Rodale crowd to make this mistake because the part of the United States they learned their stuff in (the Northeast) often has soils that hold excessive magnesium — in other words, they tend naturally to be tight and airless. Having discovered that compost corrected tight soil, the organic movement looked no further than to strongly recommend spreading both dolomite and greensand as useful adjuncts to manure and compost. Ironically, both of these rocks contain magnesium in excess proportion to the calcium they offer, thereby further tightening the same soil the compost was supposed to loosen.

During my first gardening years, I made huge compost heaps of imported materials because Rodale's *Organic Gardening and Farming* magazine instructed us gardeners to do so. We were repeatedly told there were great treasures of organic matter going to waste all around us. With a little cleverness, we could gather these things up and convert them to black gold that would grow nutritious food and make us healthy. The promise of "healthy" always inspires me. So, I bought a pick-up truck and started prowling the neighborhood on trash day, bringing home bags of the neighbors' grass clippings and fallen leaves. I had plenty of space for making compost; even better, my lot had alley access. I made some quick bins out of rigid 3-foot-high fencing wire, then I layered trash-day grass clipping collections with sawdusty horse manure, kept everything moist (wasn't easy, those bins sat in the full southern California sun, with air going in and out the sides), and let them decay. They sort of decomposed. By the second turn, it resembled compost, and it grew stuff — because I also spread oilseedmeal on the soil along with that compost. Had I not used seedmeal, my vegetables would have been pathetic.

After four years of this, I moved to rural Oregon, where I continued the attempt. However, the source and nature of compostable materials were different. The local stable had no end of material free for the hauling. It was about 75% Douglas fir sawdust and 25% horse lumps; it smelled vaguely of ammonia and urine and was only slightly damp. The first loads I brought home, I sprayed heavily with water as I heaped them up. But they didn't heat. Not at all. I tried mixing a bit of chemical nitrate fertilizer into them. No heat. After a few years, the heaps had shrunk a bit and turned a browner color, and the obvious horse turds had vanished. I called it "compost," and spread it in my garden. Fortunately, I also put seedmeal into the soil.

The Oregon garden also made crop waste that rotted slowly and wouldn't heat up in the compost heap. And the amount of finished material I got was barely (what I thought at the time was) 10% of the garden's requirement. At the time, I thought gardening was supposed to be like that.

My second year in Oregon was a wet one; the spring never seemed to settle. In June, I noticed there was a lot of baled-up hay lying on wet pastures, rotting, on offer cheaper than cheap. I brought home many loads of that stuff, stacked those soggy bales, tried mulching with them, tried composting them. What a mess! Endless grass and weeds emerged. When mixed with horse manure, the hay did decompose — slowly — but again, the growing result was poor. This stuff would have been better named "grass-straw." The spring had been so rainy that the pastures could not be mowed before the grasses had fully matured their seeds. As grass seed develops, the stems and leaves turn to straw with next to no protein in them — and almost no nitrogen. Every scarce element the grass managed to accumulate from the impoverished soils of the Oregon Coast Ranges had been put into the precious seeds. And much of that grass seed had already shattered, so at least half of it never got into the bales into the first place. No wonder that straw wouldn't heat. And no wonder the compost made from it was full of living seeds that made a huge problem. Today, I would be happy to compost rain-spoiled *green* hay made from grass that had been cut in prime condition. It would not contain half-formed seeds; in fact, to be premium stuff, the grass would need to be cut before the seed heads released their pollen. But hay-making being what it is these days, I doubt the opportunity to score such good material for next to nothing will ever come my way.

During those early Oregon years, I spread many loads of sawdusty horse manure in autumn and tilled it in to decompose over the winter. And I used lots of oilseedmeal along with it to make things grow. As I mentioned in an earlier chapter, putting that much sawdust into my land also brought with it huge quantities of potassium. I now know that the potassium imbalanced my soil and degraded my health. And the high level of imported nitrogen my soil required to perform — despite all that sawdust — also significantly lowered my food's nutrient-density. But still, I was getting a far better result than the region's compost-only gardeners.

My next playpen was 16 acres of black Malabon silty-clay loam, a highly productive alluvial soil nicely perched a few feet above the 100-year flood line of the Umpqua River. The Umpqua nation once had their permanent camp a few hundred yards from this near-ideal spot. However, by the time I got there, the land had been exhausted by a century of grain farming and a stint as a flower bulb farm. For the ten years before I bought it, the field still produced enough grass to make it worth the cost of bailing it up. When I was learning how to grow small grains on that property, I discovered that the soil was no longer fertile enough to grow cereals without fertilization, even after decomposing its thick sod.

The land was at least 50 miles away from any convenient source of organic matter. So I tilled in about a quarter-acre of sod, spread COF, and started gardening. I never had such a great result. I remember being surprised, not expecting so much success without manure or compost. However, at the finish of my garden's second summer, I could tell the garden was declining. I didn't think it was a drop in the organic matter level so much as the development of a symphylan population. The symph causes a great deal of loss and grief for most Cascadian gardeners south of Olympia, Washington, and in scattered areas north of there. The symphylan is a soil-dwelling, fast-moving, light-shy thing resembling a skinny, twisty centipede an eighth to a quarter of an inch long. It eats root tips and semi-decayed organic matter. A small population creates no observable damage. Their starting population always is small in a new garden because symphylans do not much like eating grass roots and have a hard time surviving the summers, so without irrigation they can't breed up to high levels. Symphylans often become a garden-wrecking plague after a few years, but still go unnoticed. Most gardeners conclude that some of the crop species they have are just "hard to grow." And in a way, they're right. These "difficult species," are usually the ones with roots that symphylans most prefer. The only way I know to effectively reduce symphylan populations requires eliminating watering in summer and putting the land into vegetation symphylans do not like to eat. Like pasture grasses. For a few years.

At any rate, be it the rise of symphylans, or an inevitable decline in soil organic matter, it seemed a good idea for me to shift the garden to a new spot, especially because I had six more half-acre-sized bits of old pasture readily available. And thus it was that I discovered the absolutely very best way to manage soil organic matter when food gardening. Do not make compost; import nothing but concentrated nutrition. For me, in those years, that meant COF.

Do Not Make Compost

Compost-making can be dangerous to your garden's health. It often involves bringing home material of dubious origins. There could be unwanted insects or diseased material in the stuff. If you bring in rain-spoiled hay, there will be an infinite number of weed seeds, some of them almost certainly noxious species. Because the materials going into the heap are unpredictable, the mineral content of compost is unpredictable, so the results from using it will be uncertain. I could easily add to this list.

Instead of all this travail, why not simply own enough space for four to six food gardens, but use only a small part of your space at any one time for vegetables? Put the remainder into reserve, growing mixed grasses and clovers. The reserve accumulates organic matter during a many-years-long rotation. When growing vegetables, this approach works better than any other possible method. Of course, few people have enough good land to be able to do this. But those who can, definitely should.

A serious garden, one that provides half a small family's food year-round, should, depending on climate, be at least an eighth-acre (5,000 square feet) planted in vegetables, dry beans, sweet corn, potatoes, squash or other major staples (according to the climate). Add a few smaller mouths and an aged relative or two to feed, and you're talking about a quarter-acre. But instead of having only one such garden area, I suggest having four — or six. On all but one or two of them, you should grow deeply rooting, vigorously growing perennial grasses (and some clover) and make no removals while the plot is accumulating organic matter. If you have livestock larger than chickens, feed them from other land and keep them off the future garden plots. Some months prior to the time another eighth-acre is to grow vegetables, till in the sod there (this is best done the previous autumn), and remineralize that land. Next spring, the sod will have sufficiently decomposed to allow the land to grow food crops. No compost heaps necessary.

Building Organic Matter without Making Compost

The best way to build soil organic matter is with a mixed stand of grasses, clovers and deep-rooting herbs; you could even use the more child-resistant types of lawn grasses (if they're not closely mowed), although lawn species are the least effective grasses for growing biomass. The most inspiring book I know of describing this system is

Robert Elliot's *The Clifton Park System of Farming* originally published in 1898 and reprinted in the 1940s by Faber and Faber. You can read the book online at soilandhealth.org.

I've been told of Australian graziers growing highly profitable vegetable crops as a sideline. They primarily raise livestock on high-quality acreage also suitable to vegetable production (i.e., it has a water source for irrigation). These farmers break the sod on a small part of their land and then grow one or a few crops of vegetables (or for more profit, vegetable seed) on this new ground. After harvest, the land is reseeded to mixed grasses and clovers, to be grazed for a decade or so before being called on to grow another horticultural crop. During the year that the land grew vegetables, it lost a portion of its organic matter content. During the years it grows the mixed grasses and clovers, the land rebuilds its organic matter, and purges itself of any disease organisms, insects or other minute organisms that might be interested in vegetables.

The vegetable crop that follows is a big earner. It encounters no disease problems and few insects are around that are interested, so costs of production are low. If the farmer fully remineralizes the ground before growing food crops on it, the output will be nutrient-dense. Remineralization would make the next long rest in grass even more productive of organic matter and of healthier livestock.

Sir Albert Howard praised something like this system in his last book, *The Soil and Health*. Unfortunately, few North American gardeners know his work (in part because he wrote in British English). To paraphrase Howard on this particular point: "I don't understand how gardeners think they can use an allotment for more than five years without putting it into healing grass for at least the same period of time."

If you're a typical North American, you blanked out for an instant at the word allotment and thus, missed the meaning of the statement. An allotment is a British-style community garden plot. During the Great Depression of the 1930s, British law required that any council (county or city government) make a substantial garden plot available on request to any resident of that council district. The fee was fixed at £1 per year; the plot size was, by law, 300 square yards ($^1/_{16}$ acre). The allotment system continued strong through WWII; interest in them faded away during the prosperity of the 1960s and thereafter. Recently, though, in Britain there's been an upsurge of interest in community gardens.

So what was Albert Howard taking about when he referred to "healing grass"? Vegetable crops, by themselves, throw soil out of balance; their roots secrete long-lasting chemical compounds that powerfully alter soil ecology and can interfere with the growth of following-on crops. Building far higher-than-natural levels of soil organic matter tends to promote undesirable life forms (like the symphylan), especially in a climate where the soil never freezes or gets so cold as to completely shut down the soil ecology; whatever is living in a garden can breed and breed and breed, unchecked. Diseases can find a garden and get lodged there — and have no reason to leave. Generally, five years is about the longest spell of relatively trouble-free gardening I would ever hope for in a mild-winter climate. A few years planted in grass repairs everything. Diseases and insects affecting vegetable crops have no interest in grass/clover/herb mixtures, and the root exudates of mixed grasses are nothing but positive for any vegetable crops following on. The perennial mix also restores lost organic matter.

Mowing once or twice a year and making no removals for some years is a far better approach to building high organic matter levels than grazing that grass. No matter what you may have been told by someone who supposedly knows how to maximize grass-animal efficiency, livestock — no matter how cleverly managed — do not enrich the soil. Every beast that goes out the farm gate depletes the soil by the amount of soil minerals in its body. No grazing method can compensate for that depletion, no matter how clever it is. In fact, the more clever the method, the more efficiently it will exhaust the land's mineral reserves. Worse, every beast walking around on a pasture has sharp hooves that press on the soil with great force. (Ever have a cow stand on your foot?) Especially when that pasture is moist, and the worms are active and near the surface, grazing destroys worms as it compacts the land. All things considered, grazing usually lowers the overall net production of biomass.

Grazing animals on slopes can induce a lot of erosion. I once had neighbors who had a muddy pond at the bottom of a hill that adjoined my garden. Being city folks new to the property, they did not know why so much soil was washing into it. They had been allowing someone to graze sheep on this field, thinking the sheep would save them the cost of mowing. I suggested kicking the sheep off the property and thenceforth mowing the grass once a year with a tractor, removing nothing. They did as I suggested. The next summer, the field looked like a different place. The grass was thicker; there were no more small patches of bare soil showing. After two years of no grazing and no removals, the grass stood a foot taller when in seed than it had before. And it continued to noticeably thicken and strengthen for several more years. And the pond? The water turned clear again.

If I were to idealize this method, I'd say: if you need an eighth-acre vegetable garden, then dedicate one acre to food crops. At any one time, you'll use one eighth-acre for vegetables and a second eighth-acre for perennial crops and fruit. A well-tended fruit orchard of that size should produce a gracious plenty. Many new homesteaders initially plant dozens of standard-sized fruit trees, not realizing how enormously productive a well-tended fruit tree can be — and that a person can only eat so much fruit. So, unless your intention is to grow fruit to make a lot of alcohol, you'll be better off planting fewer fruit trees and putting most of your orchard into a very few nut trees that can be given heaps of room. Heaps!

So, now we have envisioned a tidy orchard (that also houses perennials like rhubarb and asparagus), and seven potential eighth-acre vegetable gardens — only one of them actually growing food. The other six eighth-acre plots grow pasture grasses. If it is an infertile, exhausted pasture (most likely the case, because that's usually the sort of land that gets offered to homesteaders), then remineralize the entire area and replant it to a mix of grass, herb and legume species chosen to maximize the production of biomass. Mow these plots whenever the grass is forming seed. Seed heads may emerge only once in spring, but sometimes — because you mowed them — the grasses try to form seed several times in succession. You should allow all biomass to remain where you cut it. So you'll want a strong lawn mower, probably one purpose-built for this task — or you'll have to become skillful with a scythe. Once a year (best is early autumn), rotary cultivate the next-in-line of those eighth-acre plots. By spring, most of the sod you tilled under will have decomposed (assuming the soil was reasonably fertile the previous autumn), and the plot will be again be covered with tender young grasses and weeds. If the sod did not fully decompose over winter, next time spread a bit of seedmeal over the land before tilling it.

Then spread COF — or better — do a soil test, remineralize the garden, till it all in again, and when the next year rolls around, you'll get excellent, trouble-free vegetables for sure. The plot that grew veggies last year is reseeded to grasses. Yes, new gardens from sod are a bit weedy, but if you wield a sharp hoe and use wide-enough plant spacing to allow you to effectively wield that hoe, weeds will not be a problem. If you don't spread the plants out a bit and don't have a properly sharp hoe — you'll probably curse me.

There is no need to even bother to make compost with this system. Crop wastes can be spread as a vegetable-hay mulch, the remains of which will be shallowly tilled in prior to sowing the plot back to grass. Or, you can make heap compost from this garden waste along with your kitchen waste and use that for improving soil below the most sensitive, demanding crops, like celery or cauliflower. This method hugely reduces work — if you have effective, motorized tools. There are no imports of crude organic materials and no compost heaps to build, turn or spread. No diseases or unwanted insects will inadvertently be brought in. Since the future of each garden plot is, at best, from autumn one year until late spring 18 months later, there are no fancy raised beds, no enclosures, no double-digging. The garden is mostly laid out in long rows or barely raised wide beds — the old-fashioned way that adherents to "intensive methods" demean as being inefficient and wasteful. But it isn't.

So, now I have solved the whole problem for about 2% of this book's audience. Good for me!

What I actually did on my Umpqua River homestead was slightly different. The vegetable garden was one acre divided into six plots. I did till in one of them each autumn, but didn't use the new area for only one year — I used it for two. I tilled sod under in autumn. The first spring and through the remainder of that first year, I grew the most sensitive, difficult, or demanding species — cauliflower and celery, the big fancy Brassicas like Brussels sprouts and cabbages, Solanums, etc. The next year, I used that same plot to grow crops that either had the most vigorous roots or were low-demand types that grow like field crops: sweet corn, potatoes, beans (for seed), root crops like beets, carrots, kale, purple sprouting broccoli (it grows more like kale than Italian broccoli) and rutabagas. I also grew small grains in plots of about 2,000 square feet on that second-year ground. After two years of cultivation, the plot went back to grass, scheduled to remain in grass for four years before going back to vegetables for two. Unfortunately, at about the time the first plot I started on had gone around the circle of time and was about to be tilled in the second time, we sold the place and moved to British Columbia.

Right now, I am food gardening an entire quarter-acre suburban lot. All of it is in vegetables except a band around the edges growing perennial food crops (these also form a nice windbreak). I'm having lots of fun; we are helping feed half a dozen families in the neighborhood (who don't aspire to becoming remineralizing vegetableatarians). But I am now 70 years old and find myself gradually retreating from hard work. Sometimes, after an hour's hand-digging, I remember how little effort it took walking behind a well-designed, self-propelled rotary cultivator back in Oregon. I can see the day coming when I bite the bullet and buy one again, make three gardens of my quarter-acre, put two of them into pasture grasses — and let the tiller do most of the work for me. I don't know if two years in vegetables and four in grass will indefinitely maintain soil organic matter on this particular soil, but if I don't fall off the twig for another decade or two, I'll probably find out.

Buying Compost?

There are materials that substitute for homemade compost. I use them. Seems to me, if I can buy clean, effective compost or buy an industrial waste that works like compost instead of bringing in the raw materials to make it myself, if the price is reasonable and the material does not contain pests or diseases or viable weed seeds, why not? But be careful: much of what is sold as compost these days does not suit a food garden. It might be okay as a mulch under ornamentals, but not to be tilled into vegetable ground.

Clean Materials

What toxic substances can be imported with organic matter? This worry exists whenever you use pre-processed material, be it compost-like or compost, or when you import the ingredients to make your own compost. About these concerns, I do have a viewpoint to share.

The human body seems designed to withstand insults of all sorts. Some kinds of unpleasantness, like ingesting one gram of sodium cyanide at one go, are beyond the body's ability to tolerate. But otherwise, the human body is constantly being assaulted by substances it doesn't like — and it shrugs them off. Toxins are inhaled. They are naturally present in otherwise healthful foods we eat. Highly toxic substances are produced internally by mis-digestion of inappropriate foods and by the natural breakdown and replacement of our internal tissues. The body is designed to deal with these stresses; it has a powerful ability to detoxify. A well-nourished body is able to throw off an amazing amount of insult. A poorly nourished one falls off the twig at the slightest breeze.

Our fundamental health problem, the basic bottom line, is not that there are pesticide residues in our foods; the real problem is that only *residues of nutrition remain in them.* If a person's entire food intake were highly nutrient-dense, then their body would be largely unaffected by what usually comes with hidden sub-acute malnutrition. In other words, you're far better off to stop fretting over toxic traces and instead, focus on growing and eating nutrient-dense food. Our entire planet has already been poisoned by industrial and military wastes. There is no place on this planet that remains free of toxic residues. I know of no entirely, absolutely clean food. I do see sense in avoiding obvious poisoning; but I see little point in worrying about faint traces of poisons in every load of potential soil fertility — when the most dangerous poisons are being fobbed off on us in the supermarket.

On the other hand, I've been hearing of people completely wrecking their gardens by making compost with or mulching with agricultural waste containing traces of a particularly nasty herbicide. Here's a news report from the UK:

> *It is a frightful sounding tale of deformed vegetables in domestic gardens where "allotment" owners used commercially produced (non-organic) manure to supplement their soil. Gardeners have been warned not to eat home-grown vegetables contaminated by a powerful new herbicide*

that is destroying gardens and allotments across the UK. The chain of events in the UK was roughly as follows: UK farmers used a popular, commercially approved herbicide to suppress broadleaf weeds from grassland. The residues of the herbicide were absorbed into vascular tissue of grasses, where bio-degradation is slow compared to in decaying weeds and soil. Cut-grass hay containing the residual herbicide was turned into silage, and fed to cows or horses. The herbicide residues apparently did not break down in ruminant digestive tracts. Cow or horse manure (still) containing the herbicide residues was sold to domestic gardeners. (Commercial vegetable growers are not mentioned as having been specifically impacted; but it is possible.) Residual herbicide, brought in with the manure and spread in vegetable gardens or "allotments" caused deformed and/or decrepit vegetables.

The herbicide involved was aminopyralid, sold as Milestone in the US. Dow sold the same chemical as Imprelis. The obligatory legal language on the Imprelis packaging stated:

Do not use grass clippings from treated areas for mulching or compost, or allow for collection to composting facilities. Grass clippings must either be left on the treated area, or, if allowed by local yard waste regulations, disposed of in the trash. Applicators must give verbal or written notice to property owner/property managers/residents to not use grass clippings from treated turf for mulch or compost.

Dow has removed Imprelis from the market, but that doesn't mean we are through with aminopyralid. In an email exchange with me, Erica Reinheimer pointed out:

There is another version of it called clopyralid. It can go clear through a commercial composting operation, as it did in Seattle, and ruin your garden for years. What I do when I buy straw now is to soak the straw in water, then water a potted tomato plant with it for a few weeks. I water another tomato plant with a mixture from last year's straw, and if both plants are doing fine in a couple of weeks, I know the new straw is safe to use. I should do the same thing with the mushroom "compost" I am bringing in this year. It has straw in it, and mushrooms would be unaffected by a broadleaf herbicide.

Every place has its own risks and opportunities. I cannot know your situation. When it comes to locating and using potential industrial wastes as garden fertilizer, I can only make some general observations and share what I am doing now. Perhaps you'll be inspired to look around your region with new eyes. But be careful.

Tasmania's poppy fields produce much of the world's medical morphine and other opiates. (Can you imagine living in a place so tranquil that 20 acres of drug poppies are protected by nothing but an ordinary fence with a small sign saying "Prohibited, Do Not Enter"?) Tasmanian opium is extracted in an industrial manner. The field is harvested like wheat is. After the alkaloids have been extracted from the dry poppy capsules (seed pods), the residue is sold as fertilizer. Called poppy "marc," it is popular with farmers for spreading on pastures because it is high in minerals and nitrates (an analysis is provided). The worms love it, which is an excellent recommendation. When I started my current garden, I initially spread 40 cubic yards of it on a quarter-acre, which made a fluffy layer about 1½ inches thick. However, I would not spread that much a second time because another load of those minerals would throw my soil out of kilter. But one application sure woke up the soil ecology! When I lived in Oregon, there was a similar and very popular product available — mint straw, the waste product from extracting peppermint oil.

Not too far away from my town is a mushroom factory that grows its 'shrooms in large, clear plastic bags half-filled with substrate made of grain straw and chicken manure. After fruiting tapers off, they deliver these bags of "mushroom compost" to our property in quantities of 100 or more for about AU$2.50 each (the same wholesale price the local garden centers pay). Each bag holds about two cubic feet of loose, half-decomposed material, fragrant with mushroom odors. Often more 'shrooms emerge. When Annie was selling surplus vegetables out of a refrigerator by our back gate, she would put a few dozen fresh bags in a shady area and keep them moist until the fruiting stopped completely — or until I needed to spread that bag. We earned just about as much from selling those mushrooms as the compost cost, making the compost almost cost-free.

One initial spreading of poppy marc and one bag of this "compost" per square yard of bed once a year for four years, plus compost produced from my garden's own waste, brought my soil organic matter level up to 10%. Needless to say, I'm not buying-in any more mushie unless the soil organic matter level starts dropping. The mushroom compost has not fully decomposed in the bag, but after I till it in, it doesn't interfere at all with my vegetables. Seeds germinate in this soil excellently. In fact, they may sprout better with fresh 'shroom compost in the soil. I think something positive happens to the soil ecology after inoculating it with such an intense dose of mushroom spawn.

In my previous Tasmanian garden (1998–2005), I made extensive use of feedlot manure. Tasmanian feedlot cattle are bedded on wood wastes — bark and small chips — but the manure is heaped up for a year or so before being sold. At any rate, it is pretty well decomposed. However, I don't think I would have had such excellent results with this stuff had I not significantly upped the garden's nitrogen level by using a lot of seedmeal.

Beware especially of municipal compost.

The Folly of Municipal Composting

Municipal composting is supposed to be economically sensible, ecologically clean and green; therefore, all of us environmentally concerned, aware folks should support it. But is municipal compost really the greatest thing since sliced bread? I think not; the fundamental reason municipal compost is not highly desirable is the misdirected goal for making it in the first place, which is to reduce the volume of material going into landfills (this is sometimes cost-related — in some areas, a composting yard is a cheaper alternative to a landfill).

Composting the municipal waste stream is an expensive folly. The waste mostly consists of paper, cardboard and tree-trimmings from parks and roadsides (so it is mostly carbon) leavened with restaurant garbage, supermarket produce trim and sometimes sewerage sludge (as sources of nitrates). Using big equipment to make the decomposition go as fast as possible, the materials are shredded, mixed, moistened, heaped, turned, remoistened, turned, remoistened, turned, etc. This sounds like efficient industrial production. The waste heaps get quite hot, reduce down to a fraction of the starting volume, and turn black and crumbly. The product looks like compost, smells like compost... but rarely acts like good compost.

When this stuff was first put on offer south of Los Angeles, there were still many small farms and market gardens in the area. But the compost proved unpopular because it failed to make vegetable crops grow. Consequently, it had to be disposed of under roadside ornamentals or spread in parks under trees and shrubs. Then the operators of this scheme got a bit smarter and set up a parallel vermicomposting composting system. They took their "finished" compost and fed it to red worms. After the worms had digested everything they could from this material, what was left *would* grow crops and proved a popular product — what little of it was left.

Erica lives near the California coast, an hour's drive north of Santa Barbara. In an email to me she said this about municipal compost:

> *The stuff I send to the green waste is too poor to put into my compost pile. Landscapers here are reluctant to use the stuff from the local green waste. They say it can introduce undesirable weeds. Maybe true. I put all my Bermuda grass rhizomes into the green waste container.*
>
> *Those operations that do have a good starting ratio get their N from sewage sludge, which has who-knows-what-in it. What antibiotics were flushed down the toilet and ended up in the compost? Who knows? Antibiotics are not tested as a part of the compost report.*
>
> *In the US, composting operation are required to provide an analysis on request. So, you can find out the final*

C:N ratio of the finished product; for at least one batch in the past couple of years. The closer it is to 10:1 the more you might consider putting it into your garden.

In our area, there is one operation which has an excellent C:N ratio on their report, and no problems with heavy metals. That's the one that uses sewage sludge. The other one actually sent a report where their tested C:N was 20:1, and they failed coliform! And, they had heavy metal too! That's the "organic" operation that doesn't use sewage sludge. Geez! It's not that easy to buy compost around here. Finally, I decided to import shredded straw, chicken manure, and feathermeal from the mushroom composting operation nearby. It is a major pain turning the stuff myself, though.

Here's the real story on municipal composting. Huge quantities of high-carbon materials are biologically converted into much smaller volumes of mulching material that are useful for revegetation of waste sites, roadsides and ornamental beds in parks. If the decomposition process is given enough time, the tiny fraction of the starting volume that remains is useable for food production. If you consider the economics of it, the average cost of making municipal solid waste compost runs around $50/ton. When they try to sell the stuff, the usual price is either "free if you haul it" or else $1.00/ton. The only way these economics make any sense is by comparing those costs to the cost of obtaining land for dumping raw wastes on.

My concern in all this involves the economical and socially responsible management of carbon: how much carbon remains compared to what the beginning heap contained? Not much! Municipal composting converted most of the starting material into carbon dioxide gas. And much of what little nitrogen was in the starting volume was off-gassed as ammonia. Suppose, instead, that all that material had been efficiently burned, and the heat generated was used to make electricity. Wouldn't that be more sensible? That's what most European cities and towns do.

Or how about this as a far better alternative: make biochar (agricultural charcoal) with the materials. Charcoal has a high cation exchange capacity, similar to humus. And like humus, charcoal lasts hundreds of years in the soil. There are some fertile soils near the mouth of the Amazon River where a now-defunct civilization once practiced farming using biochar. The soils they built this way remain fertile — 500 years later! Making biochar involves cooking organic materials in an air-tight container, thereby releasing flammable gasses that can be cleanly and efficiently burned to heat the processing chamber generating those same gasses *and* to spin a turbine to make surplus electricity as well. A lot of electricity, in fact. One biochar cooker could power itself plus a town of 10,000 people — *and* provide enough biochar to significantly up the TCEC of the surrounding fields. (To find out more, google "terra preta.")

To evaluate the success or failure of any composting operation, including your own, apply the standards of starting dry weight against final dry weight and starting C:N to final C:N. The goal is to convert carbon into humus, not into carbon dioxide; you want to retain nitrates, not off-gas them. Why does most municipal composting do such a poor job when measured against that standard? Part of the reason is the unavoidable high C:N of their starting materials. But there's another more fundamental reason: They do not use the two most important ingredients in a compost heap — clay and rich topsoil.

If you are tempted to use municipal compost in your food garden, first ask to see the analysis. Mainly be interested in the carbon-to-nitrogen ratio, not so much in the mineral content, except to check for heavy metal contamination. If the C:N exceeds about 15:1, then for some months (or longer) after mixing it into the soil, the stuff will tie up more nutrients than it can release. It will not grow nutrient-dense food. Municipal compost can be useful as a *mulch* in your orchard or ornamental beds. Sitting on the soil's surface, high C:N materials only tie up nutrients in the surface inch as they slowly decompose.

The Folly of Excess Organic Matter

Gardeners instinctively create excesses. We practice the "moron-it system," thinking more of an otherwise good thing must be even better. The most common excesses are adding too much lime (which I discussed in the previous chapter) and adding too much organic matter.

In preparation for writing this book, I skimmed a few old compost-gardening books to identify some of the incorrect information my readers have been handed. The book that most sticks in my mind is a recent rewrite of the original 1980s book by Grace Gershuny called *Soul of the Soil*. Half the book is more or less the original 1980s Gershuny; Joe Smillie updated it to bring the book into conformity with the new organic doctrine that allows the full range of OMRI-approved substances. It is Gershuny's powerful statement of Rodale's Organic Doctrine I want to bring to your attention. Her book starts out — as so many organic gardening books do — with infinite praise for soil organic matter. Using lots of it is touted as being *the way* to grow healthy crops, feed the soil microlife, create tilth, etc. Organic matter alone, she says, improves texture and increases the soil's air supply. And if a soil doesn't hold enough moisture, organic matter is the answer. If it is heavy, airless clay, organic matter is the answer. In fact, whatever your soil's ailment, organic matter is the answer.

I know you've heard this all before.

It's a belief system that is almost a religion. And, as all religions teach a piece of the Truth (or no one would believe in them), the religion that praises organic matter as the Answer speaks a partial truth. It is true that organic matter can do all sorts of wonderful things. What is not true is that *applying heavy doses* of organic matter is the only way or even the best way to achieve those wonders. I explained in Chapter 5 that bringing the soil's balance of calcium to magnesium into a desirable zone massively improves tilth, increases air supply, and allows the soil biota to function in high gear. A soil that has calcium and magnesium in balance is able to generate its own nitrates and create lots of organic matter all by itself. When the minerals are balanced, the soil does not require heaps of compost — when you have balance, just a little dab will do 'ya.

When organic gardening books and magazine articles sing the praises of compost heaps, stand back, please, and ask yourself this: How much organic matter does the garden really need anyway? I'm here to tell you that it's a lot less than you probably thought. It's best to consider compost as food for the soil ecology and as a way to increase the soil's TCEC instead of as a source of plant nutrients (even though there are some nutrients in it). Don't think of it as a tool to massively alter your soil's texture or other mechanical properties. (The exception is for folks who are trying to grow vegetables in pure clay, which is a whole different story. But even if you suffer the misfortune of having dense clay in your vegetable garden and have been loading it up with manure and compost to improve tilth, you will probably be surprised at how less dense and sticky it becomes when its Ca:Mg is brought into balance.)

To find more complete answers, I direct you to *Factors of Soil Formation* by Hans Jenny, who was a professor in William Albrecht's department at the University of Missouri. Jenny provides a scientific explanation for the amount of soil organic matter that is really needed. When virgin land is converted from forest or prairie to farm, it usually grows great crops until its organic matter level drops too far. In other words, the quantity of organic matter a soil develops by itself at its ecological peak is the amount we should aim for. Jenny explained that if you were to measure the organic matter levels of *virgin land* along a north–south line along the Mississippi River, in steamy hot Arkansas, you would find soils with about 2.5% organic matter; around St. Louis, Missouri, you would find about 3.5%; and around St. Paul, Minnesota, about 4.5%. Given roughly the same amount of annual moisture, naturally developed organic matter levels are set more by the average temperature than anything else.

In a hot climate like California's, unirrigated, fertile land had about 1.5% organic matter before being put to the plow. When California soils are irrigated, they behave more like Arkansas soils (so, have about 2.5% organic matter). But go up the West Coast to western Washington State with year-round cool conditions, and you'll commonly find soils with 5%–6% organic matter in them. Get into northern Michigan, Wisconsin, Minnesota, much of B.C., and the well-settled parts of eastern Canada as far west as Manitoba — regions where temperatures run cool and rainfall is abundant — and it's not uncommon to find 6% organic matter. In some areas, where soil drainage was poor (slowing decomposition, but not equally retarding the production of new organic matter), the stuff accumulated in such quantity that it formed peat bogs.

First, think about what the natural level of soil organic matter would be in your region if its ecosystem were allowed to go entirely natural for a few centuries. Use that level plus 1% for your own garden's soil organic matter level target. It will fall between 3.5% in hot humid climates to about 7% in cool ones. You'll regularly be irrigating the garden if you live in a hot dry climate and so should target the 3.5% level that would occur in a hot, humid one, not the 1% or less that is the usual in desert soils. You certainly do not need 10%+, as is common in many backyard gardens located in cooler districts. I had inadvertently developed 10% at the time of my first soil test, but I never would intentionally set out to have that much. I'd have been quite content at 7%.

To develop a new garden without the guidance of a soil test, it is usually correct to assume there is not enough organic matter present unless you're turning under a highly productive hay field. Make your first action be spreading a layer of high-quality compost about one inch thick. If you're spreading half-rotted manure (and can give it time to be digested in the soil — at least from early autumn through mid-spring the next year), you can feed the soil a layer about two inches thick. Either way, compost or half-done manure, it's a one-off jumpstart. In subsequent years, your annual addition of good-quality compost could be no more than a layer ¼-inch thick; that is a gracious plenty to maintain soil organic matter at a level a good bit higher than a native state. The hardest part of achieving this is mental — it's counterintuitive to spread so little. A ¼-inch-thick layer does not completely obscure the soil. Bare patches will show through, which makes most people think they have not used enough. But if you have an established garden that

One cubic yard of high-quality compost will cover 16 beds of 100 square feet ¼-inch thick. (One cubic meter will cover those beds with six millimeters.) That's all the compost those beds will need for an entire year. So what starting volume of material do we need to end up with one cubic yard of high-grade finished compost? Can those 16 beds of 100-square-foot beds produce enough crop waste by themselves to produce one cubic yard of finished compost? Answer: it all depends.

has already received lots of organic matter, you will get *better* results if you reduce your annual applications to a ¼-inch-thick layer.

Making Powerful Compost

Results depend on juggling several factors that won't hold still. So, it's a pretty good bet that your first attempts at compost-making weren't entirely successful. You expected the heap to steam and shrink and turn to black gold, but it probably didn't. Don't despair. A non-performing heap is not a catastrophe; you can always rebuild it by adding more N, more moisture, more soil, etc. Or you can spread your unfinished compost as mulch. If your heap went the other way and got *too* hot, it just means you had too much N in the starting material. You need to remoisten the heap and turn it several times; you'll lose a good bit of N and end up without much final volume, but you can just call it a learning experience and move on to a new heap. As it's said in *The Wisdom of Solomon* (a book I've been writing since the 1970s that has now reached five pages in length): When everything goes wrong, we call it a learning experience; when everything goes right, we call it a success.

I've been making compost since 1974; this book is being written in 2012. Only in the last four years have I made excellent compost. Am I a slow learner? No. Well, maybe. In any case, it took me 35 years to realize what I was missing: *Making excellent compost requires a significant quantity of garden soil in the heap, and that soil must have some clay content.* I didn't discover this until I finally had excellent soil to work with and experienced the result from using it. Now that I know what I know, I could make good compost with almost any garden soil, so long as I had a source for good clay. So can you.

Size of the Heap

Composting is a controlled fermentation that generates heat. All organic processes are temperature related; they run faster as temperature increases — up to the point where temperature exceeds what the microlife can tolerate; any further increase of temperature works against the process. If a compost heap fails to heat up, it takes a long, long time to finish — like several years. But if it is too hot, the steaming heap off-gasses nitrates. That's absolutely the last thing you want. Let the heap get slightly hotter, and the organisms that do the actual decomposition are killed off; everything grinds to a halt until the heap cools (and gets remoistened, because heaps that get too hot also get too dry).

It's basic physics that dictates the size of a heap. The surface area of a sphere increases more slowly than its volume. And heat radiates from surfaces. If you want to cool something quickly, you spread it out and expose more surface to the air. Same with a compost heap. A larger heap encloses more volume and has relatively less surface area, therefore it retains heat better. Practically speaking, an ordinary compost heap with a starting volume less than about three cubic yards may not heat up enough except in the very center and, worse, it won't stay hot long enough. So three cubic yards is the minimum effective size. How about the other way? How large can a heap get?

Fermentation requires oxygen. Air naturally moves through a heap as long as the materials don't compact into a slimy, airless mess. A heap made from mixed food crop waste does not easily become airless. The heap's internal heat makes warm air rise and exit the top, pulling in fresh, cooler air through the base. But if the heap is too large, there can't be sufficient air exchange in its center. When that happens, microorganisms that operate without oxygen move in. Anaerobic compost is not desirable; I have a hard time even calling the gooey black stuff that comes of it compost. Practically speaking, the most workable home-garden heap is six to seven feet across at the bottom and five to six feet high (when you first build it). You can make a heap into a windrow that's as long as you wish, but no less than six to seven feet long. Smaller, it may not heat; larger, it may not breathe.

So how big a garden does it take to generate that much crop waste? My quarter-acre garden's waste, plus the trim and deadheading from Annie's roughly eighth-acre of ornamentals, makes two annual heaps; an autumn clean-up heap of about 10–12 cubic yards starting volume and a somewhat smaller spring clean-up heap (necessary because Tasmanian winters are not freezing cold, so the garden grows [slowly] all winter). Since the minimum heap size is around three cubic yards, I estimate a garden with about 1,500 square feet of actual growing beds should produce at least one heap of sufficient size at least once a year.

What if your garden is not this large? You have options. One of them is to simply abandon the idea of making really excellent compost. Look at the matter as one of convenient recycling, not as manufacture of a quality soil amendment. There are small-scale methods, such as compost tumblers, that quickly decompose smaller quantities. Or you could try vermicomposting, which does make pretty good stuff. Alternatively, you could import materials to supplement your own waste stream and make proper compost.

Containers

The question naturally arises: If I do not have enough material to make a large enough heap to heat properly, can I somehow insulate a smaller heap? Put it into a container that holds in the heat? Your answer, as usual in this chapter, is yes and no. Yes, you could make a "U"-shaped bin of highly insulating straw bales. Nothing else — just cereal grain straw. However, even straw bales restrict airflow into the heap, although not nearly as much as something solid, like wooden planks.

Composting books give a misimpression that enclosures make the process run better. This has never been my experience. However, bins and composting containers do make your yard look tidier, even if they're made of straw bales. But bins interfere with your ability to turn the heaps and, in my opinion, are a unnecessary expense (unless you make them of scrap lumber or recycled materials). Containers often prevent a heap from being heaped up high enough to hold heat when materials are in short supply because the container forces the base dimensions to be whatever the container size is. But if a heap is too short, it won't heat well.

Heat-retaining walls also reduce air flow. To overcome this, there are clever ways to build in ventilation. You can lay air-ducting or large-diameter plastic pipes with many holes drilled in them under the bottom of the heap before it is built. But I don't see the sense in first creating a problem (insufficient air due to solid walls) and then cleverly solving it, when the problem never had to exist in the first place. I advise against enclosures unless appearance is your overriding concern.

My own composting yard is a square about 25 feet on each side. In that space, I have three neat heaps; two of them are covered with a blanket of loose straw, and one is finished compost that I'm currently using. I also have an untidy "hay" stack — an ever-increasing low, spread-out pile of sun-drying garden wastes that will go into the next heap I build. Contrary to almost everything you hear, I advise you *not* to build a compost heap gradually, as materials become available. Many gardeners do it this way because backyard bin composting containers encourage it, but the decomposing process works far better and faster if the heap is constructed all at once. Then the whole thing heats up at once. So, if you can, first accumulate your materials as "hay" (which does means living with an ugly stack of drying vegetation).

My two working heaps are attractive, shaggy mounds covered with light brown straw. Loose straw not only insulates, it helps retain internal moisture while shedding rain and reducing leaching. It allows air to freely flow into and out of the heap. I urge you to cover working heaps with a thick blanket of loose grain straw. Covering each heap requires several bales. Even a foot and a half-thick layer of loose straw might not be excessive where winter is really cold. It is easy to rake loose straw off the heap when you want to turn it or start using the compost. After the straw has been in place for about a year, it loses its rigidity and starts getting compactable; what remains is ready to become an ingredient in your next heap. Even if you have composting enclosures, thickly cover the tops of the heaps with straw.

Starting C:N

The ratio of carbon to nitrogen in compostable materials varies greatly. For example, lawn clippings in late spring are extremely high in nitrogen (which means they are high in protein); even a small pile of wet grass clippings gets hot and quickly turns into a slimy mess. You can think of spring grass clippings almost like fresh animal manure. But lawn clippings at the end of summer are not much richer in nitrogen/proteins than grain straw. By the way, the best way to handle lawn clippings destined to be composted is to first spread them out thinly over the top of your stack of drying vegetation and let them become hay, thereby keeping them from heating and losing nitrogen. Or, after mowing you can let clippings cure in the sun atop the lawn for a day before raking them up. Of course, I have heaps of admiration for ex-lawns converted into food gardens.

All will go well if you rigorously avoid bringing decomposition-resistant, high-carbon materials into the heap and make the majority of the starting volume be crop waste from your vegetable garden and non-woody annual and biennial waste from your ornamental beds. Absolutely reject sawdust, bark, sticks or twigs — woody wastes of any form. You don't want anything with bark (tender flower stems and the outer leafy new growth trimmed from some hedge plants may decompose readily). Absolutely avoid paper. At one time, it was workable to compost shredded cardboard in the heap. Because the glues in cardboard were animal based, they contributed enough nitrogen to allow the cardboard to decompose readily. What glues are being used now, I do not know, but I suggest you look into that before composting cardboard boxes.

If you have high C:N material to dispose of and want to try composting it, I suggest making a separate heap with it, using twice the quantity of soil (10% by starting volume) and double-thick sprinklings of seedmeal on each layer. Expect a high-C:N heap to take at least a year to become compost, and do not plan to use it on vegetable crops, no matter how good it looks when it's finished. Mulching under ornamentals, fruit trees or other small fruit is a good use for it.

Aside from the food garden itself, the best possible sources for compostable materials are *your own* lawn, surrounds and the ornamental gardens from around your house. You can remineralize the soil growing this stuff and know it has not been contaminated. In the 70s, I used to pick up neighbors' grass clippings on trash day. I'd not do that today.

The compost quality you end up with hinges on starting with materials that contain a sufficient concentration of nutrients (in balance) with which to build the vigorous population of microorganisms that will do the actual decomposing. So it makes great sense to remineralize the entire area you're growing in. If you're a homesteader, remineralize whatever land you are mining for food-garden organic matter. When remineralizing trees, shrubs and slow-growing ornamentals, it is best to leave nitrate fertilizers out of the program because if you provoke woody ornamentals or fruit trees into the kind of rapid growth a vegetable garden demands, they may freeze out and die in winter. But there is hardly an ornamental species that does not grow better when its soil provides the full range of mineral nutrition in the same balance vegetable crops prefer. You might also have a good think about changing the sorts of ornamentals you grow; some provide more suitable compostable materials and less woody waste than others.

The best material I know of to buy-in for making compost is baled grain straw (not hay!). It's C:N will be 30–40:1. Blend two parts straw by volume to one part grass clippings (if your lawns are large, it might be best to make your annual compost heap in late spring, when the lawn is putting out the most high-protein material). Alfalfa meal makes a worthy substitute for potent spring grass clippings, especially the loose stuff that accumulates around the stack of alfalfa bales at your local feed and grain dealer. I have been allowed to sweep that stuff up and haul it away, no charge. How much alfalfa to how much straw? Around two parts straw *by weight* to one part alfalfa. Even better, mix straw and alfalfa half-and-half with dried vegetable-garden crop wastes.

If yours is a suburban family with a big lawn, consider how to make high-quality food garden compost. You probably have no source of fresh animal manure, do not keep chickens or rabbits or other homestead livestock, and are disinterested in humanure. The lawn substitutes. During summer, garden waste is accumulated, spread out in thin layers on top of the previous layer of waste, making a hay stack drying in the sun. Because you're going to restack that dry vegetation when converting it into a compost heap, it's a wise practice to make sure nothing going into that haystack is more than one foot long. Corn and sunflower stalks should be first cut into short pieces, huge broccoli plants or Brussels sprout stalks chopped into foot-long lengths before being set out to dry. Lawn clippings and ornamental waste can also be spread atop. So, too, can a reasonable quantity of autumn leaves, although leaves tend to have rather high C:N and also tend to pack tightly, making a heap airless. They should not make up too much of the heap or be concentrated into one layer, but blended throughout. If you have great quantities of autumn leaves, it's helpful to run them through a hammermill while they're dry, or run a lawnmower with a bagger over them. This reduces the volume by about two-thirds. Store the chopped, dry leaves under cover (I put them in old feed sacks) until it is time to marry them into a compost heap. Accumulation of compostable materials can continue into winter in a mild climate. Stacking materials in a windrow to dry means the various materials are layered from bottom to top, but when you make the heap, you remove materials from one end, thus every layer of the heap you're building gets roughly the same mix of vegetation. When spring grass mowing begins, the actual compost heap is built by layering a year's accumulation of dried vegetation with fresh grass clippings and a bit of seedmeal or COF, and always, soil. I gauge how much COF to use by spreading it on each eight-inch-thick layer of dry vegetation about as thickly as I'd spread it on soil. If your remineralized lawns are several times more extensive than your food garden, and you're using spring grass clippings to activate the heap, then boosting the amount of nitrogen with seedmeal or COF should not be necessary. If turned once in midsummer, the new heap probably will be close enough to finished by autumn that it could be spread and shallowly dug in to blend itself into the soil over winter. Or better, perhaps, turn the heap at summer's end, let it continue to work over the winter, and use that thoroughly mature compost in spring.

We have a small patio for socializing, but no lawn. Our place is designed to feed us first and please the neighbors last. So I have no grass clippings. I use seedmeal instead. In a new garden, I suggest using COF to heat the heaps because it takes more than nitrogen to sustain a large microbial population. Crop wastes from soil that has not been fully remineralized do not contain high-enough mineral levels. Keep that foremost in mind if you're buying-in materials with which to make compost. Garbage in; garbage out.

Clay and the Nature of Humus

If you garden on sandy soil, without doing something a bit unusual, you will not be able to make high-quality compost — even if you do everything exactly as I suggest and even if you mix garden soil into the heap when it is being built. When your heap has turned itself into something resembling soil, you may think you've made a heap of humus. But it's more likely that what you have is a pile of half-decomposed organic matter, not humus. Its rate of decomposition has slowed, and the most easily consumed parts have been eaten. What remains looks like humus, but it is not humus. When you put that black crumbly stuff into warm soil, the material continues to rot, and it will do this fast enough to provoke a good deal of plant growth. As this material continues to decompose in your soil, a fraction of it *may* turn into humus *if* the soil contains clay. Otherwise, it'll rot away to nothing.

Humus formation requires the presence of clay. Although the following allusion is not entirely correct or complete, imagine that humus is created in the gut of an earthworm. There, its digestive juices cause the clay in the soil passing through its gut to combine chemically with whatever organic matter the worm's gut is not able to digest. The result is humus. But the nature of that humus depends on the nature of the clay in the soil. If you're on sandy soil, the miniscule quantity of clay in it probably has a low cation exchange capacity. Consequently, most of your compost will not convert into humus, and what little bit of humus you do end up with will not have a high CEC. If you're on a highly developed, geologically old (weathered) clay soil, the humus formed will not have the highest possible CEC either, but at least you'll get humus. The CEC of humus can vary from about 100 to about 400. High-CEC humus only forms from high-CEC clay. And when it comes to most of the miraculous things humus does for soil, 100-CEC humus is a quarter as effective as 400-CEC humus.

If your sandy soil has a clay subsoil (and many do, in temperate humid climates), the affordable way to get clay into the heap is to dig a small pit and mine some. Turn a bucket of subsoil into a bucket of clay soup with an electric-drill-driven paint mixer. Dip a small broom into the slurry and do a thorough sprinkle of clay over each layer of your heap as it is being built. You end up with a lot more compost from the same mass of starting material. If your garden does not feature a clay subsoil, maybe there's a road cut where you can pinch a bucket or two of clay. How much clay? About 1%–2% by starting volume. So, a 4-cubic yard starting-volume heap (27 cubic feet per cubic yard) needs about two 5-gallon buckets two-thirds full of clay.

Soil in the Heap

To make humus, the heap must include rich garden soil. Hopefully, the soil contains a fair bit of high-CEC clay. Soil performs several crucial functions; the lack of it explains why municipal composting gets such poor results.

Soil is the natural home of microorganisms that convert ammonia gas into nitrates. These organisms only live in soil. Ammonia-converting microbes allow farmers to inject pure ammonia gas into damp soil and have next to none of it escape. All the injected ammonia dissolves into the soil moisture where it is (almost instantly) microbially converted to ammonium cations that adhere to clay. During the composting process, decomposing proteins release their nitrogen content as ammonia. If this gas is instantly captured by the soil, it is not lost to the atmosphere. So, if your heap smells at all like ammonia, you're losing a lot of value.

To end up with powerful compost, the heap must burn off carbon until the C:N gets down to around 12:1. Then you can GROW stuff with it. If the starting materials in a heap have an average C:N of 36:1, and your finished compost ends up at 12:1, then you've burned off, or eliminated, two-thirds of your starting carbon to get there (in the form of carbon dioxide lost to the atmosphere). Hopefully you've lost none of the nitrogen the heap began with. However, if the heap is losing nitrogen in the form of ammonia as it is off-gassing all that carbon, then the 12:1 ratio is not achieved until even more carbon is burnt off. So, you might end up with only a quarter the starting volume — or even less than that — by the time the heap finally settles at 12:1. So how much soil is needed to retain the ammonia? About 5% of starting volume. To get that 5% and to have it thoroughly blended into the heap, when I build a new heap, I sprinkle a thin layer of my best garden soil over each 8-to-12-inch-thick layer of crop waste.

If your garden soil is clayey, your need for clay in the heap is taken care of. If it's a loam soil, only 5% by starting volume should still supply plenty of clay. If you have a very light, coarse-textured loam with a low clay content (by definition, loam contains 10%–30% clay), or a sand or silt soil almost entirely lacking in clay, you need to add some clay. I'd use a broom to spray some clay slurry between each layer, as well as including garden soil.

Soil serves to slow down a compost heap, something like the moderating rods in a uranium nuclear reactor. This is highly useful because there is a lot more ammonia lost when the temperature goes too high. If your feedstocks contain too much nitrogen for the amount of carbon in them, you'll find that your heaps get too hot. In that case, mix more soil into them; try up to 10% soil by starting volume. In the same way, if you are composting pure livestock manure, mix about 10% soil by starting volume into the fresh manure as you heap it up. If you have a runaway heap, tear it apart and rebuild it, mixing in more soil as you do. Not a huge amount more. There is a huge difference in performance between having 5% soil and 10% by starting volume.

Garden soil serves to mass-inoculate the heap so that fermentation begins immediately and the heap heats quickly. Even if you tried, it is almost inconceivable that you could build a new heap entirely lacking the necessary decomposers (unless the entire starting volume was sterile paper and cooked food wastes). But, like any other ferment, it is important to encourage the life forms you want to take over promptly, avoiding the possibility that the ferment will go the wrong way. This is much like making alcohol: you first sterilize the sugary water and then inoculate it with a strong yeast culture of the exact strain you desire. Otherwise, you risk making vinegar or off-tastes in the alcohol. Same with compost.

I mentioned how municipalities composting high C:N materials without soil compound their folly. But you don't have to. If you lack the proper in-puts, there are commercial compost inoculants that provide free-living, ammonia-fixing bacteria that prevent loss of N without having to use soil in the heap for that purpose. Sometimes, these same inoculants provide phosphate-liberating bacteria, etc. But as useful as inoculants may be, they do not provide clay.

Anions in the Heap

Clay does not attract and hold anions; in fact, it repels them. So you need the anion exchange capacity of humus to prevent borate, nitrate, phosphate and sulfate from leaching. If the soil itself has a decent organic matter content, it is unlikely that anions in solution will remain in solution very long before being assimilated by the microlife.

Soil microorganisms constantly release nutrients as well as assimilate them because they are steadily dying and decomposing, releasing mineral nutrients that are taken up again by other microlife or by plants.

However, phosphate is another story. This anion has a strong tendency to form highly insoluble calcium or iron phosphates. As MAP dissolves, or as an OMRI-approved fertilizer releases phosphate, this anion doesn't hang around for too many weeks before becoming insoluble. There's but a short window of opportunity when plants can uptake phosphate fertilizer.

To keep phosphate available for years instead of for weeks, and to get much more response from each phosphorus dollar you spend, first incorporate it into your compost heap. I know two basic approaches: either the phosphate fertilizer goes into the heap's starting volume and goes through the entire composting process (the best method); or else the fertilizer is blended into finished compost during the final turn-out and allowed to merge into the material for a month before it is spread. The way I gauge how much fertilizer to put into compost it is to reckon how much area the finished compost will cover against how much phosphorus I want to spread over that area.

As I write these words, I have a small heap of finished compost — about one cubic yard — awaiting use. There's enough in that heap to thinly cover about 10 beds of 100 square feet each. As I was turning this compost out into a loose pile so I could spread it easily, I mixed into it 55 pounds of soft rock phosphate containing five pounds of actual phosphorus. The SRP is now merging into the humus' anion exchange capacity and is being incorporated into the bodies of its active microecology. When this fortified compost gets spread a quarter-inch thick, the application rate will be about 175–200 pounds P per acre (and the rate of application of compost will be about five tons per acre).

About the same final SRP concentration could be put into a heap while it is being built. To gauge how much, assume that the final volume will be between 33% and 40% of the heap's starting volume, and then compute how much area that final volume will cover. My heaps usually are built in about six layers, so I would spread about one-sixth of the fertilizer on top of the sprinkling of soil I put over each layer.

A highly knowledgeable homesteader named John Slack proudly told me about how he makes mineralized compost. His soil was brought into balance years back, and he is now aware of which minerals his soil is capable of maintaining out of its own deep reserves and which minerals (and quantities) have to be regularly supplied. These go into the compost.

There's another good reason to add minerals to compost heaps. Imported raw materials were probably not grown on balanced soil, so they won't contain the highest possible levels of plant nutrients. Thus, they will not do the best possible job of feeding the compost heap's ecology. In short: a heap built with nutrient-undense materials will not heat as fast or make highly mineralized compost. Perhaps your own garden soil has not yet been brought into balance. In that case, you would be wise to make up something like the Complete Organic Fertilizer I describe in Chapter 4; sprinkle that as generously on each layer as if you were fertilizing a growing bed. That much COF in your compost won't be nearly enough to run your entire garden, but it will make your compost much better.

Kitchen Garbage

One of the worst plagues a food garden could suffer is a flock of English sparrows (aka "flying rats") nesting in the immediate neighborhood. People who carelessly feed backyard chickens often attract sparrows. Cereal-based kitchen wastes such as old rice or stale bread interest sparrows. Cooked foods of all sorts appeal to rats and mice. Scattering these to dry out atop the growing stack of drying vegetable garden waste hay may not be such a good idea if you're living in suburbia.

Vermicomposting

One way to compost kitchen wastes, while effectively keeping vermin out of them, is to use a covered worm bin. Most plastic composting enclosures (holding one or two cubic yards) are actually vermicomposters in disguise. They are not large enough to heat up for long, especially so when they are gradually filled with new material as the old stuff settles. If your composting bin is open to the soil at the bottom (some municipalities disallow this), you don't necessarily need to import worms to get things started. Red worms will almost certainly be present (in small quantities) in your soil. If they can get to it, the worms will soon enough discover this rich source of food. As long as the contents don't get too hot, the bin will soon be filled with red wrigglers. Once they get established, it's amazing how quickly food wastes and grass clippings disappear.

Vermicomposting is especially suited to climates with a mild winter; red wigglers cannot survive freezing. While researching my aforementioned composting guide, I found a suggestion to move the family worm farm into the basement over winter. In order to find out for sure if it would be possible to live in close proximity to indoor worm composting, I put a worm box under the kitchen sink, displacing the usual wastebasket and supplies. After a few weeks, the kitchen developed a slightly fruity/vinegary aroma, especially noticeable when the under-sink door was opened. We could have lived with that. Had the worm farm in the kitchen become important enough, I could have vented the under-sink cabinet to the outside with a quiet computer fan. However, after ignoring the slight fragrance for a week, we then discovered a dozen different varieties of insectoidal life forms prospecting on the kitchen counters, their numbers diminishing by their distance from the worm bin. Thus, the worms got promptly exiled to the garage. I suppose a bin could be kept going during a hard winter if it were placed in an unheated mud room or porch, so long as it never actually froze in there, but red worms are not active in chilly soil, so the worm bin wouldn't accept much feeding until spring. I apologize for being a big vague about worms and winter, but it has been 50 years since I lived somewhere where winter meant frozen soil.

Making Compost in the Soil

Tasmania is still a somewhat backward place; it only got modernized in the 1960s. To handle kitchen scraps, country and small-town Tasmanians used to dig a few feet of shallow ditch between long, well-separated rows of vegetables. They buried their kitchen scraps therein, covered them with a few inches of soil, and then advanced the ditch with each load of kitchen waste. Really old-timey rural Tasmanians without a sewerage system or running water (they mostly used rainwater-capture tanks), used a dunny for sanitary purposes. (A dunny is a 5-gallon metal can kept in a small outside building resembling an outhouse; the structure is also "the dunny.") Far less odorous than a typical "longdrop" outhouse, every few days, the dunny can or pail would be emptied into the same ditch the kitchen scraps went into. When the garden was spaded up the next spring, rows of buried, decomposed organic matter became rows of vegetables, and the rows where vegetables had been the last year became the place to bury the garbage and humanure.

A similar large-scale system, termed *sheet composting*, was once highly recommended by Rodale. Raw manure and other organic wastes were spread atop the ground and rotary cultivated. Lime and phosphate rock could be spread at the same time. Months would pass (usually autumn and winter); in spring, the ground would be tilled again and planted. These days, certification bureaucrats restrict this practice for reasons of public so-called health — with some added justifications about how decomposing raw manure can off-flavor the food being grown. I see it this way: it's mighty sad when average health has declined to the point that people become fatally ill from exposure to a little animal shit.

Making Quality Compost

Making low-grade compost is easy. Nearly all organic gardeners do it, which is why I have had such success promoting the use of Complete Organic Fertilizer. The major lack in most home-garden compost is nitrogen. This deficiency almost always happens because the decomposition process doesn't go far enough. The heap may heat and cool, and the material can look like compost, but the C:N isn't yet 12:1. It'll be more like 20:1. So when this pseudo-compost is mixed into soil, it does not release an abundance of plant nutrients. I apologize for being negative; TV presenters of gardening information are always positive, smiling, eager and enthusiastic. But in the case of making compost, if you don't do it right, it don't come out right. And then it don't grow things right. And then they don't nourish you right. Right!

The main obstacle to making good compost is the slow-motion learning curve. Most gardens only generate enough waste to make one or two heaps a year. Someone gardening where there is wintry winter may start only one heap each year. In my mild climate, heaps still require an entire year to finish. I guess that over ten years, the average gardener will, at best, have the opportunity to make 20 compost heaps.

When I was a younger man still possessed of a strong liver, I used to enjoy homebrew. My own soon became so good I had to fend off beer-swilling visitors. But the first batch I made ended up being discarded. The fifth batch was drinkable, but not as good as store-bought. The tenth was more than drinkable, and it was far better'n Bud. When I became a slightly older man with a somewhat weakened liver, I got into making bread using fresh wholemeal flour milled on my own kitchen countertop. My first batch turned rock hard as soon as it cooled. Fortunately, almost anything made of flour and water is delicious when still hot out of the oven, especially if you melt some butter into it. My sixth batch didn't turn hard when it cooled down, and it tasted okay, but it was crumbly, like cake. So was the tenth.

Then I found out a few things about wheat — about its variable protein content, that virtually all wheat protein is gluten, and that there is rampant ignorance amongst the folks at the health food store. They insisted only two kinds of wheat existed: organically grown and conventional. They had no idea of the protein content of either sort of wheat berry they sold and had no interest in doing the hard work required to source an effective bread-making wheat berry. So I started buying "conventional" wheat from a local Seventh Day Adventist lady selling flour mills and baking supplies. She actually cared about protein and knew how to make good bread. After making a few batches with proper ingredients, I had the basics pretty well worked out, and soon could depend on an excellent result. But there was an inevitable learning curve. Thing of it was, I could make a new batch of bread every few days, so I learned how to do it in a few months. Brewing took longer to master than baking did; each batch fermented for a few weeks before it was ready for bottling. To smooth it out, it needed to rest in the bottle for at least three months. And each batch, 5 gallons at 8%, was plenty of beer to last a few weeks, even with many freeloading guests. Consequently, it took longer to gain skill at beer making than it did to learn how to bake good bread.

Beer and bread are the easier sort of ferments. Once you have sourced the proper ingredients, the process is repeatable and the outcome predictable because the ingredients are fairly standardized. Switch to a new harvest of bread wheat berries, and your dough may slightly alter its nature, but only a little. It doesn't take long to adjust. Compost is also a fermented product. However, to experience ten heaps may take a gardener ten years. The ingredients going into most compost heaps are unpredictable. They often are whatever was readily available at the time. So the ways in which this constantly changing stream of ingredients interacts in the compost heap are not reliably predictable.

But I have a bit of experience at this game, so forthwith, here is a summary of what I've learned:

- *Size:* The heap must hold heat, but the core must breathe. So the heap needs to be at least six feet across at the base and no more than seven (and seven is twice as good as six); it must be at least five feet high at the start, and no higher than six feet; it must be at least six or seven feet long. If you lack materials to make a heap large enough to work effectively, but you live where winter doesn't freeze everything solid, get a big, continuous-feed plastic composter with a lid and make vermicompost.
- *Air supply:* The foot-thick bottom layer of the heap really should be foot-long pieces of corn stalk, sunflower stalks, Brussels sprout stalks, or the like, spread irregularly so they won't pack tight; these allow fresh air into the bottom of the heap to replace the warm air rising out the top. If you do not have an uncompactable bottom layer in place, you're probably going to have to turn the heap every few months or build in some clever air-ducting. In my own practice, I make the main autumn clean-up heap at the same time I'm bringing in the old corn stalks. These, chopped into foot-long pieces form an open bottom layer. And for the spring heap, it's Brussels sprout stalks and tough old broccoli plants on the bottom.
- *Moisture:* When building the heap, water each layer well before starting the next. If you get the entire heap moist while building it and then thickly insulate the heap with loose straw, you probably won't have to turn the heap to add any more water. A thick straw blanket can save you a heap of work.

- *Extra nitrogen:* If you do not have your own source of fresh manure to layer into the heap as it is being built, then abundantly sprinkle each layer with seedmeal or, better, with complete organic fertilizer. If you can stockpile bottles of urine in the garage or toolshed, I suggest doing that. Pour a gallon into each layer as you build the heap. Old urine will give the heap a bit of aroma, but only for a few days.
- *Materials:* Please believe me! To end up with effective compost, you must not put woody material or paper into the heap. Your starting C:N must not exceed 30–35:1. If the starting materials are not at least half food crop waste or trim from annual/biennial flowers grown on fertile ground, you won't get the best compost. If you can source nitrogen-rich materials, like low-grade alfalfa, pea straw, mint straw, etc., you might make these up to one-third the starting volume in place of animal manure. However, these have a tendency to compact and become airless. Don't use too much. Most of the heap's starting volume must come from the garden itself. If you must buy-in materials, cereal grain straw is the closest thing you can conveniently buy that has a C:N similar to mixed garden trim and waste.
- *Insulation:* Covering the heap with a foot-thick (or thicker) layer of loose grain straw is critically important.
- *Location:* Where you put your compost heap has a lot to do with core temperature and loss of moisture. In summer, it's best to compost out of the sun. Close to the shady side of a building is a good spot. Under a tree is not a good spot; the tree roots may steal a lot of value and dry out your heap. But putting a barrier under the heap to keep out tree roots also prevents worms from entering (and leaving).

  In the cool or cold season, wind protection can be important. An old shed or garage with a dirt floor and at least three crude walls to break the wind is an ideal place for a heap that has to go through a freezing winter; the shade of a roof would help during the high heat of summer. On the other hand, a clear roof during winter would make the structure into a semi-greenhouse, possibly effective enough to prevent a heap from freezing solid.
- *Turning:* I can't predict how often you'll need to turn your heaps. My location, my materials, my methods, require one turn, halfway through, and a final "turn-out" that loosens the compost and

prepares it for spreading. If a heap gets dry, it needs turning, and you need to spray it with a lot of water while so doing. If the heap smells of ammonia, it needs turning, watering *and* more soil. If it cools, turning and remoistening it may cause it to heat up again. But high heat is not necessary; as long as the heap's core is moist and getting sufficient air, there is no absolute need to turn, unless you're in a hurry for it to finish.

- *Temperature (and duration)*: There's a lot of confusing information about the temperature to strive for. Temperature exceeding about 155°F makes the organisms of decomposition die off, so 155°F is the peak core temperature. Many experts say to bring the heap to about 150°, close to peak speed. However, the microorganisms that convert ammonia gas die off around 140–145°, and that's when the heap usually starts smelling like a horse barn. You don't want that! I suggest the maximum temperature you ever want to see is about 135°. A medium-heat heap takes longer, yes. But it makes far more finished compost. And that compost will have a far more favorable C:N. The quick-easy way to take the heap's core temperature is to push a sharpened stake or stick about four feet long into the heap and leave it there. When pulled out you, can feel the stick's temperature. Then put it back in. If a heap fails to get hot enough, next time add more nitrogen, more manure, more COF. Perhaps make it larger. Use a thicker insulating straw blanket. If a heap gets too hot, tear it apart, add more soil and remoisten; next time, use less nitrate fertilizer or manure in it.

The best book ever on how to make compost was written by Sir Albert Howard; it's *The Waste Products of Agriculture*, published about 1932. (I scanned it and put it on soilandhealth.org for free download.) *Waste Products* will show you not just how to make strong compost, but how to turn a compost heap into a nitrate-production factory. Normally, much of the starting nitrogen in a heap is lost to the atmosphere. Using Howard's method, you can build a heap with 100 pounds of nitrogen in the starting mass and not only retain all that you began with, but end up with 120 pounds of nitrogen in the finished compost. The additional nitrogen is manufactured by the heap's ecology during an artfully managed fermentation. To accomplish this agronomic miracle, however, Howard relied on large numbers of desperately poor laborers. Still, the book is illuminating and will help you make far better compost.

Don't forget, should the heap heat for a few weeks and then cool down, there may be no huge reason to turn it to force it to heat up again. Why be in a rush? Let it work slowly for a few months. Let that straw blanket work for you; keeping the core temperature even 20 degrees warmer than the average ambient air temperature will double the speed at which it decomposes. The best compost takes at least six months; a year's time is even better.

Chapter 10

Epilogue

This book is only a bare-bones beginning. You now possess a system that permits you to analyze soil and produce a great growing result without knowing the full science of soil fertility. Success with this system requires only careful obedience and good arithmetic. Erica and I spent much time and energy on the system's fine points in order to eliminate the possibility of you making major errors. We know our readers are beginners. Our main consideration has been "safety first." To a practicing soil analyst, my system will seem unnecessarily cumbersome. But it works.

There's an inevitable leap in understanding that only comes after you've analyzed a few dozen soils. If you manage to pull in soil samples from thousands of miles around (like I did when writing this book), you'll come to see how the broad patterns of soil fertility work. If you analyze a few dozen samples from close by, you'll soon see the similarities and differences in the soils in your region. Either way you'll get smarter.

I hope to speed your progress by alerting you to a few risks and sharing a major short-cut. You'd learn about these in any case, but perhaps I can prevent some mistakes and save some advisee of yours from not getting the full result — or a bad result. And reader be warned: In the rest of this chapter, I am not addressing novices. I'm speaking to someone who has already analyzed a few audits and is considering becoming a practicing neighborhood soil analyst.

Excess Calcium and the Analysis

Working out the prescription for a *heavy soil* that has not been fertilized or limed previously is the easiest sort of exercise. The great majority of these soils have multiple deficiencies but no significant excesses. Add fertilizers in roughly the correct quantities (or as much fertilizer as my system's application limits permit), and the crops will grow excellently as the soil moves toward the targets. In a few years of repeating the analysis and fertilizing accordingly, the soil will come into balance. If liming is done cautiously, using fine-grind lime and done per soil test results, the soil need not contain much unreleased limestone.

Light soil requires delicate handling. The soil's exchange capacity cannot hold sufficient plant nutrients to grow even one crop. It is wise to assume light soil will not adequately feed most of the trace elements. It can't possibly hold enough potassium, and, if a light soil's organic matter level is not high, there may be problems maintaining a supply of the anions. Solutions include increasing the exchange capacity by abundantly spreading compost, side-dressing when growth slows (with slow-releasing materials when possible), and split applications. As the soil becomes ever-lighter (i.e., its TCEC goes below 7), the use of a truly Complete (and balanced) Organic Fertilizer becomes ever-more essential. Soils managed this way are almost certain to accumulate some free lime.

Being aware of the presence of free lime is crucial when doing a Mehlich 3 soil analysis, especially when doing an analysis for a home gardener. The most frequent difficulty arising while *testing* garden soils comes from the previous (mis-) or (over-) use of lime. "Excess" calcium will do no major harm — other than to degrade a soil test's accuracy. If you're a bit clever about handling free lime that shows up on the audit as available calcium, you can make seat-of-the-pants adjustments to the soil report without going to the bother of retesting, and you often won't have to bother getting an elevated pH ammonium acetate extraction test for the bases.

The most important soil test is the one done *before* any fertilization or liming happens. You do it to make a record of the native soil's starting condition. And what you want to know most from this initial audit is the TCEC *before any free lime gets involved*. If you have limed and fertilized but not done such a test, it may not be too late. In that case, look for a bit of similar soil nearby that has not been much amended (a lawn, perhaps). Test that spot to establish a baseline TCEC.

Free lime can massively increase the number representing the soil's exchange capacity on a soil audit, but free lime does not increase the functional TCEC. I have seen highly calcareous desert soils with an actual TCEC of 10 or 12 be mis-assigned a TCEC of 42 on an M3 audit; in Chapter 8, I included a real-life audit done on a heavily limed sandy loam soil. Although its actual TCEC was around 10, it was reported at 34.79! As an exercise, I suggest you put the unadjusted numbers from that audit (for "Dave," seen in Figure 8.5) into the *Acid Soil Worksheet* (found in the Appendix) and see what it looks like.

You should see *apparently* large magnesium and potassium deficits and a big excess of calcium. If you amended the soil with the 456 pounds of magnesium and 901 pounds of potassium that seem to be called for, you'd do a lot of harm as well as waste a lot of money. That's the main reason the *Acid Soil Worksheet* carries application limits on magnesium and potassium.

In the future, remember there are a few things that will instantly alert you to the presence of enough free lime to significantly distort the soil audit: a pH over 7.0; calcium saturation over 70% (in Dave's case, Logan reported 88%); the apparent need for large quantities of Mg and K; and an obviously overstated TCEC (obvious because Dave's soil is a sandy, and sands do not develop a TCEC much above 10.0). Fortunately, Dave had tested his soil before spreading so much ag lime, and it reported a TCEC of about 10.0. Seeing the TCEC leap from 10 to about 34 alerted Dave that there was something wrong.

Almost all gardeners spread manure and/or compost, so it is a reasonable guess that Dave's soil now has an organic matter content that is somewhat higher than it was originally. Organic matter at 4.12% is not particularly high, but it probably is higher than it was at the start. If the organic matter had been lifted to 7% from a starting TCEC a few years ago of 10.0, then you might expect to see actual TCEC raised to 14.0 from 10.0, but not to 34.79. Seven percent organic matter is probably about as high as a sandy soil in Dave's climate could practically be brought to, so a TCEC of 14.0 wouldn't be a bad high-end guess for Dave's sandy loam. So, let's assume the TCEC is 14.0, and we are going bring that soil back to an Albrechtian target of 68% calcium saturation. Provisionally, plug in a TCEC of 14.0 and recalculate the targets for calcium and magnesium at 68:12, then look at the potassium level my system calls for at 14.0.

Calcium target: 400 x 14.0 = 5,600 x 0.68 = 3,808 lb/ac
(compared to 12,296 lb/ac "found" on the audit).

Magnesium target: 240 x 14.0 = 3,360 x 0.12 = 403 lb/ac
(compared to 1,001 target on the original audit).

Potassium target from the chart: = 365 lb/ac
(compared to 1,085 target on the original audit).

Looked at through Albrechtian spectacles, Dave's soil has a lot of free lime, a surplus of magnesium and a large shortfall of potassium, but not nearly as large as it seemed if we thought the TCEC was 34.

You'll never go seriously wrong when you calculate a soil prescription based on a TCEC that is lower than it actually may be. The worst that'll happen is that you'll fail to add enough of some element to make a difference. Or you may incorrectly conclude there's enough of a trace element when there really is a deficit. But these are easy things to fix — next time. But had Dave spread magnesium at the rate called for by a soil with a TCEC of 34 (403 lb/ac), the soil would probably have tightened up considerably. And then he'd need to spend several years trying to leach that magnesium out with gypsum. Had Dave ignored the potassium application limit on the *Acid Soil Worksheet* (200 lb/ac) and applied all 901 lb/ac of potassium that seem called for by a TCEC of 34.79, he could have pushed the pH up to 7.6. Or higher. (Because the even-more excessive magnesium and huge surplus of potassium would knock some calcium cations off the exchange points.) This elevated pH would not have a desirable effect on nutrient availability. Having such high levels of Mg and K might also induce functional deficiencies in calcium. Not to mention the money wasted.

Anytime you suspect there is a great deal of free lime present, it's wise practice to reduce the TCEC calculated by the soil lab. How much? That's a matter of experience. There are no naturally calcareous soils in my state. When I discover a local garden with a pH over 7.0 and a calcium saturation higher than 68%, I know it's due to the gardener's previous liming. I do not bother with a fizz test or with testing the bases using an elevated pH ammonium acetate extraction; what I do is first reduce the amount of discovered calcium by one-third (sometimes by one-half in extreme cases) and then recalculate the TCEC (thereby lowering it to roughly two-thirds of its original value because most of the calculated TCEC comes from the calcium component). This simple adjustment produces an entirely different picture — an effective, workable picture.

Mike Kraidy gave me a rule-of-thumb on handling highly calcareous *desert* soils given a Mehlich 3 audit. Note that Mike's approach does not work for calcareous soils in humid climates; these can be truly heavy soils with a big clay content and a genuinely high TCEC. But desert soils usually don't have much clay content. They are naturally light soils or, at most, barely over the light/heavy line. By shifting a few numbers, Kraidy's method allows you to convert the M3 audit of a naturally light calcareous soil into something useful. No matter what the audit says about calcium saturation, no matter what it says about the TCEC, simply reduce the amount of calcium discovered to 6,000 pounds in the furrowslice acre (or 3,000 ppm), recalculate the TCEC accordingly, and then use the *Calcareous Soil Worksheet* (saturation targets 85% calcium, 5% magnesium). The usual outcome of this manipulation is to see that calcium saturation now falls a few percentage points short of 85%; you fill that calcium shortfall by adding gypsum (which also helps leach excess magnesium, potassium and/or sodium). If the pH is much in excess of 8.2, there's almost inevitably going to be excess magnesium and/or potassium after the TCEC has been recomputed.

When Things Go Wrong

Sometimes a soil prescription doesn't work; the garden fails to grow well, or fails to grow better than previously. Usually it's because there are two invisible, yet powerful fertility-altering factors at work. Best that you are aware of them. One comes from above, the other from below. I'm referring to the *water* and *the subsoil.*

The Water

Rainwater is pure. It may contain traces of nitrates or sulfur but otherwise, it is naturally distilled water. Irrigation water can be different. If it comes from a pond filled only by surface runoff, it'll be practically free of dissolved minerals. But if it comes from underground, from springs or wells, it may contain high levels of dissolved minerals, especially magnesium and sometimes, sodium. The concentration of these minerals can be so high that they overwhelm whatever is in the soil.

If you're looking at a soil audit showing surplus magnesium or sodium, *and if it is irrigated soil,* the first thing to check is the water. If you've been trying to balance a soil containing excesses, and the excesses don't seem to be lessening, check the water. If the water holds a lot of dissolved minerals, your only remedy may be to calculate how many pounds of magnesium or sodium the irrigation water is contributing, then put in sufficient gypsum to provide enough sulfate to combine with whatever amount of sodium and/or magnesium you are dealing with. That could mean several tons of gypsum per acre per year, ongoing. That could also mean a consultation with an experienced soil analyst.

The Subsoil

My own garden is easily diggable until you get down about 12 inches. Then the soil color shifts from brown to red, getting ever redder, and holding ever-more clay content as you go ever deeper. The soil is still a clay loam from 12 to 24 inches, just a bit more clayey. This holds true down to at least six feet — clay content increasing with depth, but not a pure clay soil. This red clayey subsoil strongly resists shovel, spade and fork. And root. Some vegetable species are capable of getting into that subsoil; some find it difficult or impossible. There's a chemical reason: my soil test for 12–24 inches shows that the subsoil has a magnesium saturation of 24%, but the calcium saturation is only 35%. (The test results are shown in Figure 10.1.)

My subsoil is also potassium deficient (1.2% saturation) and contains a large reserve of sulfur (270 ppm). There's almost no zinc or copper, and manganese is rather low.

From this soil/subsoil audit, I can draw several conclusions about the garden's ongoing maintenance:

My plants are never going to be short magnesium, especially if I can reduce the subsoil's magnesium saturation and increase its calcium saturation, thereby opening it to greater root penetration. This

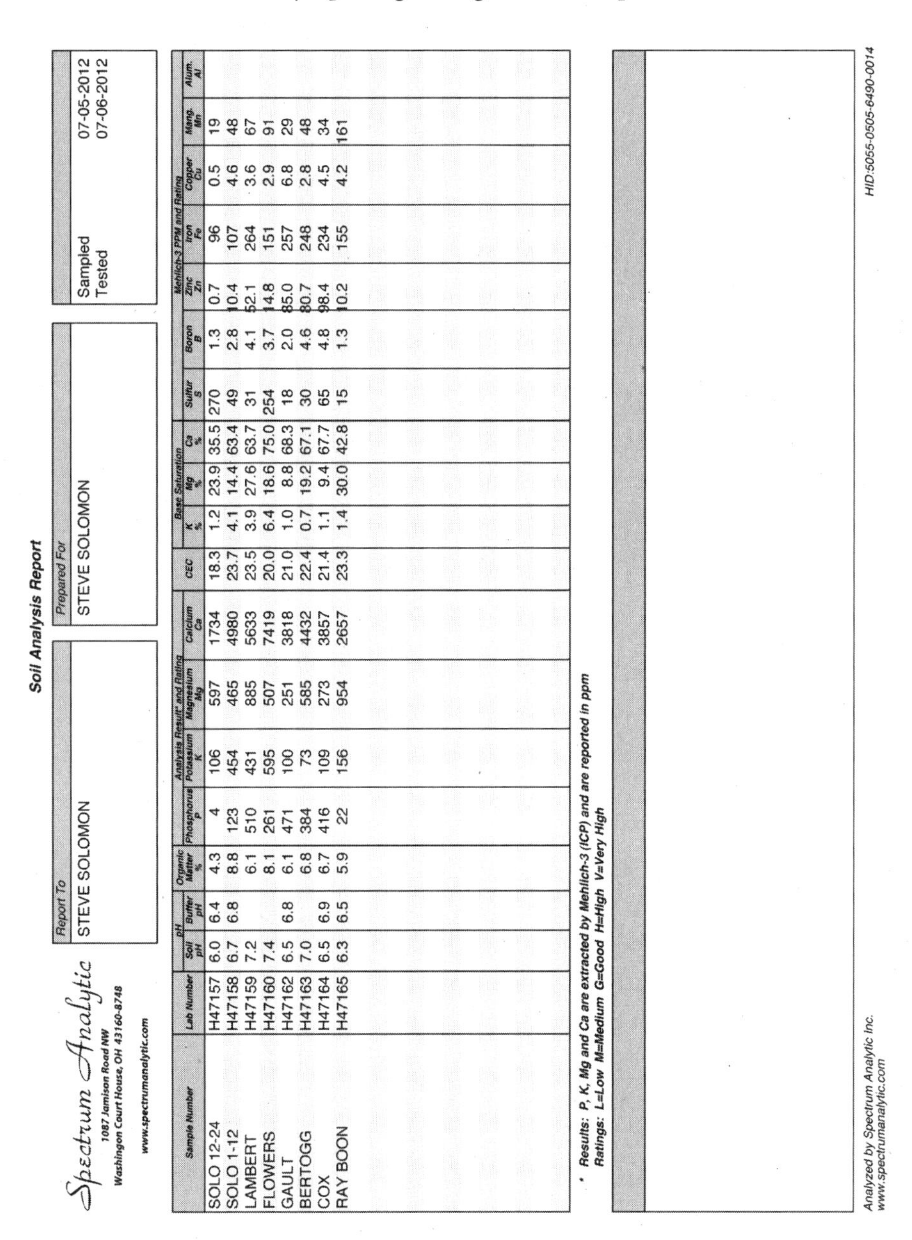

Spectrum Analytic
1087 Jamison Road NW
Washington Court House, OH 43160-8748
www.spectrumanalytic.com

Soil Analysis Report

Report To: STEVE SOLOMON
Prepared For: STEVE SOLOMON
Sampled 07-05-2012
Tested 07-06-2012

Sample Number	Lab Number	pH: Soil pH	pH: Buffer pH	Organic Matter %	Analysis Result* and Rating: Phosphorus P	Potassium K	Magnesium Mg	Calcium Ca	CEC	Base Saturation: K %	Mg %	Ca %	Mehlich-3 PPM and Rating: Sulfur S	Boron B	Zinc Zn	Iron Fe	Copper Cu	Mang. Mn	Alum. Al
SOLO 12-24	H47157	6.0	6.4	4.3	4	106	597	1734	18.3	1.2	23.9	35.5	270	1.3	0.7	96	0.5	19	
SOLO 1-12	H47158	6.7	6.8	8.8	123	454	465	4980	23.7	4.1	14.4	63.4	49	2.8	10.4	107	4.6	48	
LAMBERT	H47159	7.2		6.1	510	431	885	5633	23.5	3.9	27.6	63.7	31	4.1	52.1	264	3.6	67	
FLOWERS	H47160	7.4		8.1	261	595	507	7419	20.0	6.4	18.6	75.0	254	3.7	14.8	151	2.9	91	
GAULT	H47162	6.5	6.8	6.1	471	100	251	3818	21.0	1.0	8.8	68.3	18	2.0	85.0	257	6.8	29	
BERTOGG	H47163	7.0		6.8	384	73	585	4432	22.4	0.7	19.2	67.1	30	4.6	80.7	248	2.8	48	
COX	H47164	6.5	6.9	6.7	416	109	273	3857	21.4	1.1	9.4	67.7	65	4.8	98.4	234	4.5	34	
RAY BOON	H47165	6.3	6.5	5.9	22	156	954	2657	23.3	1.4	30.0	42.8	15	1.3	10.2	155	4.2	161	

* Results: P, K, Mg and Ca are extracted by Mehlich-3 (ICP) and are reported in ppm
Ratings: L=Low M=Medium G=Good H=High V=Very High

Analyzed by Spectrum Analytic Inc.
www.spectrumanalytic.com

HID:5055-0505-6490-0014

Fig. 10.1.

should be more than possible. Besides, the subsoil can't be all that hostile; I know there already is some root penetration because the subsoil organic matter level is 4.3%. My topsoil also has excess magnesium — 14.4% saturation. Therefore, magnesium-containing fertilizers of any kind are to be strongly avoided.

My subsoil is never going to serve as a source of potassium. I probably should invest in a few hundred pounds of potassium sulfate and keep it in the garage because I'll know I'll have an ongoing need for it. Similarly, the topsoil is going to need ongoing supplementation with zinc and copper. If I live long enough, the quarter-acre garden will probably eat a 25-kg sack of zinc sulfate and half a bag of copper. The subsoil is also very short phosphorus. I am already planning on spreading 175 lb/ac phosphorus per year for quite a few years; some of it will lodge in the subsoil. I hope.

To accelerate these subsoil shifts, I am going to spread about 1 ton/ac gypsum per year, and continue that for the next few years before retesting the subsoil. If those additions of gypsum plus the calcium in the soft rock phosphate do not lower the magnesium saturation of my topsoil and simultaneously move the calcium saturation close to 68%, then in a few years I'll be spreading more ag lime.

As I mentioned earlier in this chapter, it takes working out a dozen or so soil audits before you start grasping the patterns. So, below the audit of my own soil, are a handful of local gardeners' soil tests I had done at the same time. Have a study.

Contemplating these audits will expand your mind. Contemplating a few more for your neighbors, friends and family will expand your abilities. Soon, some of you will be practicing neighborhood soil analysts. As this happens, more people are going to awaken to the sad results of the industrial food system.

And maybe we will have a brighter future because of that. I was married to Dr. Isabelle Moser for 15 years. Isabelle was a practicing naturopath who prescribed water fasting to heal serious illnesses. Some of her clients had mental or emotional difficulties as well as physical complaints. Dr. Moser was a real doctor; she had a PhD in psychology. Her early practice focused on treating schizophrenics with dietary reform, megavitamins, exercise and detoxification. Isabelle said that trying to help an unhappy person with talk-therapy was slow, cumbersome and ineffective. It was far easier to repair their bodies. As the person's overall health improved, complex, agonizing mental and emotional concerns vanished. They became entirely irrelevant.

In the same way, I see that many of our current social problems would also vanish by themselves, if only the mass average health of people were uplifted. It is my belief that this could be accomplished — even without the conscious or intentional involvement of most people. All we would have to do is grow our industrial food crops with the goal of making them as nutrient-dense as possible. Then, in order to be extremely well, bright, vigorous and happy, the average person would only have to make reasonably healthy choices most of the time.

This book has given you the skills needed to begin that social transformation — one neighbor, friend or family member at a time. I hope you will give it a go.

Appendix A: Sources

Soil Mapping

If you want to find out what your soil type is and get some information about its agricultural or mechanical potentials, look up your land up here at this website: websoilsurvey.nrcs.usda.gov/app/HomePage.htm.

Finding a Soil Analyst

To locate a local analyst that specializes in helping gardeners and homesteaders, or to register yourself as an analyst, go to soilanalyst.org.

Buying Fertilizers

- North Carolina State University has a list of OMRI-approved material suppliers, ces.ncsu.edu.
- Seven Springs Farms. 426 Jerry Lane NE, Check, VA 24072, (800) 540-9181, 7springs@swva.net; 7springsfarm.com/catalog.htm.
- Black Lake Organic. 4711 Black Lake Boulevard, Southwest Olympia, WA 98512, (360) 786-0537, info@blacklakeorganic.net, blacklakeorganic.com.
- Down To Earth Distributors. Eugene, Oregon (800) 234-5932, down-to-earth.com. Contact them to locate retail outlets for their fertilizers.
- Concentrates Inc. 5505 SE International Way, Milwaukie OR 97222, (503) 234-7501, 800-388-4870, concentratesnw.com.
- Peaceful Valley Farm Supply. 125 Clydesdale Court, Grass Valley, CA 95945, (888) 7841722, helpdesk@groworganic.com.

Computer Soil Analysis

To obtain a copy of the Reinheimer spreadsheet, go to www.grow abundant.com.

Sending Soil Samples to the USA

The American government's soil import system checks all soil samples at their point of entry to insure that they are going to a soil lab equipped to maintain what Plant Protection and Quarantine (USDA-PPQ) and Homeland Security term "biosecurity." At the point of entry, an employee working for one of those two agencies will check the import documents and clear the parcel for delivery.

Everything I say about this procedure is up-to-date as of this book's publication; however, governments being governments, it would be wise to check with the soil lab in advance of sending samples to make sure there have been no significant changes.

1. To save postage, I suggest that you air-dry and then sieve your soil sample so it contains no twigs, stones, etc., only clean, fine, dry soil. Send at least 70 grams (about 2.5 ounces). Put the soil into a small plastic zipper bag and seal it carefully. Write your surname or other brief identifier (should be less than 10 characters long) on a strip of paper and tape that to the outside of the plastic bag. PPQ requires that the soil be double-bagged to make sure it does not leak out. So, put the labeled bag inside another, similar zipper bag and seal it also. If you have several samples to send, each one must be double-bagged. When I send a batch of samples, I put all those double-bagged samples into yet another larger, sealed plastic bag; I hope my overcaution makes those grumpy quarantine officials smile slightly.
2. Download the lab's *sample submission form*. Print out two copies. (For Logan Labs, go to loganlabs.com/doc/HowToFillOutLogan LabsWorksheet.pdf.) Fill both copies out. (If you are new to soil analysis, I strongly urge you to use Logan Labs. If you are an experienced soil analyst you'll save a few dollars per sample by using Spectrum. In Chapter 10, you'll find some soil audits done for me by Spectrum; have a good look and see if you feel comfortable using that report.)

3. If you're using Logan Labs, download their *soil import permit* from their website; www.loganlabs.com/doc/soilpermit.pdf. Print out two copies; the permit is several pages long; include the entire permit, twice. (If you want to discourage yourself, read all the fine print.) Otherwise, simply notice that Logan's permit number is located near the top right of the first page. If you're using Spectrum Analytic, contact them and request a copy of their import permit be sent to you by email.
3. Place one copy of the lab's filled-in sample transmittal form and one copy of the soil import permit in a business-size envelope. Seal it.
4. Photocopy and cut out the PPQ form (below). Don't forget to neatly print the permit number in the space provided on the PPQ-550.

Fig. A1.

PERMIT NO. ....

This Package Contains

PLANT QUARANTINE MATERIAL

DELIVER TO

U.S. DEPARTMENT OF AGRICULTURE

ANIMAL AND PLANT HEALTH INSPECTION SERVICE
PLANT PROTECTION AND QUARANTINE PROGRAMS

PPQ Form 508 (2-87) *Previous edition may be used.*

U.S. DEPARTMENT OF AGRICULTURE
ANIMAL AND PLANT HEALTH INSPECTION SERVICE
PLANT PROTECTION AND QUARANTINE
4700 RIVER RD., UNIT 136
RIVERDALE, MD 20737-1236

SOIL SAMPLES
RESTRICTED ENTRY

The material contained in this package is imported under authority of the Federal Plant Pest Act of May 23, 1957.

For release without treatment if addressee is currently listed as approved by Plant Protection and Quarantine.

PPQ FORM 550
(MAR 95)

Tape or glue the form to outside of the business-size envelope. Also boldly mark this envelope: "IMPORT PERMIT ENCLOSED."

5. You need a strong shipping envelope or box to hold the soil sample(s). It must be large enough that you can affix the business-size envelope to the front of it and still leave plenty of room for the lab's delivery address. I use a prepaid postal document express pouch with a 1 kg limit; I can send a dozen samples and all the paperwork within that 1 kg limit. Any seams on this larger envelope that could potentially leak soil must be taped over.
6. Address the large envelope to the soil lab. Your local post office should accept the package because you have included the proper import permits.

To repeat: Inside the smaller envelope (that will be affixed to the larger one) go one copy of the lab's import permit and one copy of the lab's transmittal form. The PPQ-550 form is affixed to the outside of the smaller envelope. The smaller envelope is affixed to the shipping envelope (box).

What you now should have ready for posting is an addressed shipping envelope (or box) that has its seams covered with tape (so no soil particle could possibly escape); to its front is taped a business-size envelope with "IMPORT PERMIT ENCLOSED" written on it and a form PPQ-550 taped on it. Inside the larger envelope (or box) are the second copies of the import permit and lab transmittal forms and the double-bagged soil sample(s).

Should work.

FAQ

All my previous books were about things I had known for many years. This book sits at the cutting-edge of my personal knowledge. Thus it is incomplete. I have never explained this subject before; thus the book must be weak in some respects. So Erica and I are going to create a Frequently Asked Questions on soilanalyst.org. If anything in this book seems hard to grasp, please respond by contributing your confusion, suggestion or objection to the FAQ. Thank you.

Appendix B: Bibliography

*Books marked with an asterisk are available for free download from the Soil and Health Library, at soilandhealth.org. Most of them are in the "Ag Lib Collection." Some of these titles can be downloaded from the Soil Analyst website, soilanalyst.org.

*Albrecht, William A. *Soil Fertility and Animal Health*. Webster City, IA: Fred Hahne Printing Co., 1958. This was also published in 1975 by Acres, U.S.A. as *The Albrecht Papers: Volume II*. It's a must-read!

Astera, Michael. *The Ideal Soil*. Published by the author, 2010. Obtain a copy from Michael's website, soilminerals.com. The book is also distributed by Acres, U.S.A. I started writing *The Intelligent Gardener* in order to explain Michael Astera's system to a broader audience than a self-published book can usually reach and perhaps, being an ex-schoolteacher, I hoped I could explain the method better. Along the way, my book developed into something beyond *The Ideal Soil*. Well worth a read.

Coleby, Pat. *Natural Farming: A Practical Guide*. Scribe Publications: Carlton, VIC, 2004.

*Dale, Tom and Vernon Gill Carter. *Topsoil and Civilization*. University of Oklahoma Press, 1955. This classic survey of world history demonstrates how every civilization from Mesopotamia to Rome has destroyed its agricultural resource base and thus destroyed itself. The book also looks at modern-day Europe and the United States, concluding there is considerable uncertainty about the sustainability of our own system.

*Elliot, Robert. *The Clifton Park System Of Farming*. London: Faber & Faber, 1943. Reprint of the 1898 original. Originally published as *Agricultural*

Changes, this book's thesis was broadened by Sir Albert Howard, Newman Turner, Louis Bromfield, etc. Elliot developed a system of laying down land to grass, dependent on little input but a complex mixture of deep-rooting pasture seeds. The pasture rotations would be broken after four to eight years, row crops grown until the humus levels declined to a threatening level, and then the field would be restored to grass/clover/herbal mixtures.

*Ernle, Lord. *English Farming Past and Present.* fifth edition, London: Longmans, Green & Co., Ltd. 1936.

Foth, Henry D. and Boyd G. Ellis. *Soil Fertility.* New York: John Wiley & Sons, 1988. A comprehensible, brief university-level text for serious soil analysts in training. Available online for free download at soilanalyst.org.

*Hopkins, Cyril. *The Story of the Soil.* Boston: Richard G. Badger, 1910. One of the best "made-simple" holistic soil manuals ever written, all wrapped up as a romance about a bright young man with a solid ag-school education going out to buy a farm and falling in love. WARNING: This book expresses views on race that in its day were considered normal and acceptable but in our day are viewed as shockingly racist. Those who cannot view such expressions as "historical documents," should not read *The Story of the Soil.*

*Hopkins, Donald P. *Chemicals, Humus and the Soil.* Brooklyn, NY: Chemical Publishing Company, 1948. Hopkins makes the point that chemical fertilizers are effective and positive *to the degree that humus remains in the soil.* He felt that the real problem with the use of chemicals was the suggestion that chemicals could replace farmyard manure. Hopkins takes on the Howardites point by point and demolishes many of their positions. The book's arguments are cogent and largely correct, although Hopkins's "scientific" biases distort his objectivity in areas relating to human health. This book should be carefully read by anyone who considers themselves "organic."

*Howard, Albert, *An Agricultural Testament.* London: Oxford University Press, 1943. Howard wrote this book for the general public with the intention of creating a new form of farming.

*Howard, Albert and Yeshwant D. Wad. *The Waste Products Of Agriculture: Their Utilization As Humus.* London: Oxford University Press, 1931. Probably Howard's most important scientific publication, detailing the nature, practice and significance of Indore composting, especially as it pertained to Indian agriculture.

*Howard, Albert. *Farming and Gardening for Health or Disease.* London: Faber and Faber, 1945. Published in the United States as *The Soil and Health: A Study of Organic Agriculture.* Chronicles Howard's life history and outlines the complete breadth of his contribution.

Jackson, Carlton. *J.I. Rodale: Apostle of Nonconformity*. New York: Pyramid Books, 1974.

Jenkins, Joseph. *The Humanure Handbook: A Guide to Composting Human Manure*. Published by the author; available on Amazon.com.

*Jenny, Hans. *Factors of Soil Formation: A System of Quantitative Pedology*. New York: Dover, 1994. One of the most important books about soil ever written. It should be thoroughly studied by anyone seeking a full understanding of soil fertility and how to handle agricultural soils. This is a scientific text that can be understood without high-level mathematics, however, a well-grasped secondary school science education and a touch of geology will go a long way toward making this book fully comprehensible.

Kinsey, Neal and Charles Walters. *Hands-On Agronomy*. Acres U.S.A.,1993.

*Krasil'nikov, N.A. *Soil Microorganisms and Higher Plants*. Moscow: Academy of Sciences of the USSR, 1958. Translated by Dr. Y. Halperin. Printed in the United States by the Government Printing Office. This is the ultimate study of the microbial process in soil. In the Soviet Union of the 30s, 40s and 50s, industrial production was scanty. Had Soviet agronomic research focused on increasing yields through the use of voluminously spread chemicals, the result would have been massive crop failures because Soviet industry could not have produced chemical fertilizers and pesticides in large enough quantity. So Krasil'nikov focused on the biological process, and he found ways to improve plant growth by crop rotation and the production of special composts and microbial ferments of the sort that could be produced in an old barrel by a farmer. All these "primitive" solutions are based on a very high-level understanding of the microbial process in soil and the interactions among soil microbes with each other, of how crop species interact with each other via long-lasting soil residues (root exudates), and how plants and microbes interact with each other.

Lovel, Hugh. *A Biodynamic Farm: For Growing Wholesome Food*. Austin, TX: Acres, U.S.A., 2000.

McKibben, William. *The Art of Balancing Soil Nutrients: A Practical Guide to the Interpretation of Soil Tests*. Austin, TX: Acres, U.S.A., 2012.

Parnes, Robert. *Fertile Soil: A Growers Guide to Organic and Inorganic Fertilizers*. Davis, CA: Ag Access, 1990.

*Pieters, Adrian J. *Green Manuring Principles and Practice*. New York: John Wiley & Sons, 1927. A thorough review of all known information about green manuring and its contribution to the maintenance of soil fertility and the improvement of agricultural productivity. Replete with tables and photographs of great historical interest.

*Price, Weston A. *Nutrition and Physical Degeneration.* New York: Paul B. Hoeber, Inc., 1939. Currently in print in paperback from the Price-Pottenger Nutrition Foundation. In the 1930s, Dr. Price traveled to isolated regions finding people who, because of their proper nutrition, enjoyed general good health, long life and virtual immunity to dental disease. Price traveled to remote parts of Scotland, Switzerland, Canada, Alaska, Peru, Africa, Down Under, etc. He found similarly healthy (and isolated) peoples who did not partake of the industrial food system. The book contains remarkable photographs that show the comparison between what healthy bodies and physically degenerated bodies look like — far better than words ever could. No one who spends time studying these pictures will ever view the health and appearance of their friends, neighbors — or their own face in the mirror — in the same way.

*Rodale, J.I. *Pay Dirt: Farming & Gardening With Composts.* New York: Devin-Adair, 1946. A collection of pro-humus-farming and gardening odds and ends, mostly from early *Organic Gardening* magazines.

*Rodale, J.I. *The Organic Front.* Emmaus, PA: Rodale Press, 1948. Tens of thousands were swept up by the intense enthusiasm of J.I. Rodale at the inception of the American organic gardening and farming movement. Almost immediately, there developed intensely polarized antagonism between the innocent "organicist" and the technologically proficient "chemicalist." Hostilities persisted at least into the 1980s and perhaps even longer. Some of the causes of this conflict occurred because J.I. strongly and directly opposed powerful economic interests, but still, a great deal of this hostility may have been created by J.I. Rodale's own attitudes. *The Organic Front* will be very interesting to anyone seeking to understand the history and personalities involved in the organic gardening and farming movement. Most of this book consists of articles from early issues of *Organic Gardening* magazine.

*Rodale, J.I. *The Healthy Hunza.* Emmaus, PA: Rodale Press, 1949. J.I. gathered, recounted and evaluated all available data at the time for this research project.

Smillie, Joe and Grace Gershuny. *Soul of the Soil: A Soil-Building-Guide for Master Gardeners and Farmers.* fourth edition, White River Junction, VT: Chelsea Green Publishing Co., 1999.

*Smith, J. Russell. *Tree Crops: A Permanent Agriculture.* New York: Harcourt & Brace, 1929. Nutrient-dense food the easy, sustainable way — without plowing, soil erosion, fertilizing. A classic that should be read by everyone.

*Tiedjens, Victor A. *More Food From Soil Science: The Natural Chemistry of Lime in Agriculture.* New York: Exposition Press, 1965.

Walters, Charles, ed. *The Albrecht Papers.* four volumes. Kansas City, MO: Acres, U.S.A., 1975. Vol. 2 of *The Albrecht Papers* is a reprint of Albrecht's only actual book, *Soil Fertility and Animal Health,* which he self-published in 1955.

*Weaver, John E. *Root Development of Field Crops.* New York: McGraw-Hill, 1926. Chapter I contains what may be the best basic soil manual there is; Chapter III suggests magnificent realizations about how to grow plants with an awareness of their root activities and how that affects what one experiences above-ground. Anyone intending to grow plants well needs to *study* both of Weaver's books, especially the first portions of this one. Of interest to organic growers will be Weaver's frequent citation of Albert Howard's researches in India.

*Weaver, John E. and William Bruner. *Root Development of Vegetable Crops.* New York: McGraw-Hill, 1927. The classic study is filled with species-by-species illustrations, each worth tens of thousands of words to someone who wants to grow vegetables better.

*Wrench, G.T. *The Wheel of Health.* London: C.W. Daniel Company Ltd., 1938. (Reprinted in 1960 by the Lee Foundation for Nutritional Research, and again in 1990 by Bernard Jensen International, Escondido, CA.) This small book is Dr. Wrench's classic exploration of the Hunza, a mountain people renowned for their longevity and vigor. It should rest at the very foundation of one's personal explorations of health and its roots. The book includes a summary of the lifeworks of two other renowned health "explorers," Sir Robert McCarrison and Sir Albert Howard. Dr. Wrench was an individual possessed of a most admirable intelligence.

Appendix C: Worksheets

Acid Soil Worksheet
Excess Cations Worksheet
Calcareous Soil Worksheet.

There are three worksheets, each one is two pages long. Make as many copies as you need. I suggest photocopying them back-to-back, so the entire worksheet is on a single sheet of paper. The full-size worksheets are available as a free download at:

tinyurl.com/IntelligentGardener or **SoilAnalyst.org.**

If I am inspired to adjust these worksheets, the latest version will be downloadable.

Acid Soil Worksheet

U.S. Measurements

Name ______________________

Plot or Field ______________________

Date of Test ______________________

Sample Depth 6 inches All numbers on this worksheet assume a six inch sample depth

TCEC __________ **TEC:** If CEC is below 10 it is a "light soil." Over 10 is "heavy soil."

pH __________ **pH:** If pH is 7.0 to 7.6, go to the Excess Cations Worksheet. If pH exceeds 7.6, you may have calcareous soil.

Organic Matter % __________ **OM:** Target over 7% in cool climates. South of the Mason-Dixon Line target over 4%. Assume an approximate release of 15–25 lb nitrogen per 1% OM. Varies with temperature, moisture and soil air supply. $N = 0.22 \times NO_3$

Element	Actual Level	Calculating Target Level (Pounds per acre)	Target (Pounds per acre)	Deficit (Pounds per acre)
Sulfur **S**	ppm lb/ac	**S = ½ Mg (Target Level)** until there are no cation excesses; then **you *may* Target S=1/3 K**		
Phosphorus **P**	P_2O_5 P =	**P = K (Target Level)** Calculate using actual P, not phosphate. $P = 0.44 \times P_2O_5$		
Calcium **Ca**	lb/ac	**TCEC x 400 x 0.68 = Target Level**	Minimum target 1,900 lb/ac	
Magnesium **Mg**	lb/ac	**TCEC x 240 x 0.12 = Target Level**		
Potassium **K**	lb/ac	**K is proportional to TCEC: see chart**		
Sodium **Na**	lb/ac	**TCEC x 460 x 0.02 = Target Level** Be certain of good water quality before adding sodium	Do not exceed 160 pounds	
Boron **B**	ppm lb/ac	**B = 2 lb/ac if CEC below 10** **= 4 lb/ac if CEC above 10**	Do not exceed 4 pounds	
Iron **Fe**	ppm lb/ac	**Fe = 100 lb/ac if CEC below 10** **= 150 lb/ac if CEC above 10**		
Manganese **Mn**	ppm lb/ac	**Mn = 55 lb/ac if CEC below 10** **= 100 lb/ac if CEC above 10**		
Copper **Cu**	ppm lb/ac	**Cu = ½ Zn (Target Level)**		
Zinc **Zn**	ppm lb/ac	**Zn = 1/10 P (Target Level)**		

Potassium Target Levels

TCEC	Pounds	TCEC	Pounds	TCEC	Pounds
		16	390	28	493
		17	400	29	500
		18	410	30	507
7	255	19	420	31	511
8	270	20	435	32	515
9	290	21	443	33	519
10	310	22	451	34	523
11	320	23	459	35	527
12	335	24	463	36	531
13	350	25	475	37	535
14	365	26	481	38	539
15	380	27	487	39	543

	One acre, six inches deep weighs	One hectare, 80 mm deep weighs
1 meq Calcium	**400 lb**	**400 kg**
1 meq Magnesium	**240 lb**	**240 kg**
1 meq Potassium	**780 lb**	**780 kg**
1 meq Sodium	**460 lb**	**460 kg**

1 ppm = 1mg/kg = 2 pounds/acre = 2.24 kg/hectare

If TCEC is lower than 7, use value for 7. If it is over 39, use value for 39.

Date of Issue: 07/07/2012

Download full-size worksheet free at tinyurl.com/IntelligentGardener or soilanalyst.org.

Acid Soil Worksheet, page 2

	Deficit From other side of worksheet	**Application Limit** Per acre/year	**Quantity and Material to Add**	**S**	**Mg**	**Ca**
Sulfur **S**		110 lb 90% Ag Sulfur				
Phosphorus **P**		175 lb/ac elemental P				
Calcium **Ca**						
Magnesium **Mg**		No more than 10% of target magnesium per year				
Potassium **K**		200 lb/ac elemental K				
Sodium **Na**						
Boron **B**		2 lb/ac elemental B				
Iron **Fe**						
Manganese **Mn**						
Copper **Cu**		No more than 7 lb elemental Cu				
Zinc **Zn**		No more than 14 lb elemental Zn				

	N	P	K	S	Ca	Mg
Fish Bone	4	8.8		.06	19.0	.03
Fish Meal	10	2		0.6	2.3	.03
Crab Shell	3	1.5	.025	.02	23.0	1.3
Blood Meal	13	0.5				
Feather Meal	12	0.0	0.35	0.4	0.6	
Bone Meal****	3	13.0		2.5	12.0	0.3
Oilseed Meal	6	1.5	1.0			
Copra Meal	4	1	0.7			
Kelp Meal	1	0.3	2.5	2	2	0.7
Ag Lime					32-39	2
Dolomite					22	13
Gypsum				17	20.5	
Oyster Shell					36	0.03
Magnesium Oxide						50
Montana Hard Rock Phos**		1.3			29	
Calphos		8.8			20	
Monoammonium Phosphate		23 (Plus 12% N as NH_3)				
K-Mag			18.2	22		11
Langbeinite			15.6	23		12
Greensand***		.05	6	1.3	1.5-3.0	2-4
Ag Sulphur				90		

Sea Salt	35% Sodium (Na)	
Borax	10% Boron (B)	
Iron Sulfate	18% S	30% Fe
Manganese sulfate	19% S	32% Mn
Copper Sulfate	12.5% S	25% Cu
Zinc Sulfate*	17% S	35% Zn
Potassium Sulfate	17% S	42% K
Magnesium Sulfate	13% S	10% Mg

* Zinc sulfate picks up moisture from the air; store in airtight container.

** Hard Rock Phosphate is 1.5% available P and contains around 27% insoluble phosphate.

*** Greensand contains 9% Fe, 50% Si and many trace elements. More than half its potassium content is insoluble.

**** Bonemeal contains 5.7% sodium.

Download full-size worksheet free at tinyurl.com/IntelligentGardener or soilanalyst.org.

Excess Cations Worksheet

U.S. Measurements

Name ______________________

Plot or Field ______________________

Date of Test ______________________

Sample Depth 6 inches All numbers on this worksheet assume a six inch sample depth

TCEC ____________ The Mehlich 3 extraction overstates TCEC when free lime is present and should not be used if *the soil reacts when doing the Fizz Test*

pH ____________ This worksheet is for M3 audits on non-calcareous soils with a pH of above 7.0. If pH is over 7.6, do a Fizz Test. High pH can be caused by high levels of calcium, magnesium and/or potassium and/or sodium.

Organic Matter % ____________ Organic matter levels exceeding 5% are extremely helpful. From normal soil organic matter decomposition, assume approximate release of N = 15–25 lb N per 1% OM. Varies with temperature, moisture and soil air supply. N = 0.22 x NO_3

Element	Actual Level	Calculating Target Level Pounds per acre	Target Pounds per acre	Deficit
Sulfur **S**	ppm lb/ac	**S = ½ Mg (Target Level)** until cation excesses eliminated; then switch to the Acid Soil Worksheet		
Phosphorus **P**	P_2O_5 P =	**P = K (Target Level)** Calculate using actual P, not phosphate. P = 0.44 x P_2O_5		
Calcium **Ca**	lb/ac	**TCEC x 400 x 0.68 = Target Level**		
Magnesium **Mg**	lb/ac	**TCEC x 240 x 0.12 = Target Level** If deficit less than half of Target Level, do not add Mg		
Potassium **K**	lb/ac	**K is proportional to TCEC: see chart**		
Sodium **Na**	lb/ac	**TCEC x 460 x 0.01 = Target Level** Be certain of good water quality before adding sodium		
Boron **B**	ppm lb/ac	**B = 2 lb/ac if CEC below 10** **= 4 lb/ac if CEC above 10**	Do not exceed 4 pounds	
Iron **Fe**	ppm lb/ac	**Fe = 100 lb/ac if CEC below 10** **= 150 lb/ac if CEC above 10**		
Manganese **Mn**	ppm lb/ac	**Mn = 55 lb/ac if CEC below 10** **= 100 lb/ac if CEC above 10**		
Copper **Cu**	ppm lb/ac	**Cu = ½ Zn (Target Level)**		
Zinc **Zn**	ppm lb/ac	**Zn = 1/10 P (Target Level)**		

Potassium Target Levels

TCEC	Pounds	TCEC	Pounds	TCEC	Pounds
		16	390	28	493
		17	400	29	500
		18	410	30	507
7	255	19	420	31	511
8	270	20	435	32	515
9	290	21	443	33	519
10	310	22	451	34	523
11	320	23	459	35	527
12	335	24	463	36	531
13	350	25	475	37	535
14	365	26	481	38	539
15	380	27	487	39	543

	One acre, six inches deep weighs	One hectare, 80 mm deep weighs
1 meq Calcium	**400 lb**	**400 kg**
1 meq Magnesium	**240 lb**	**240 kg**
1 meq Potassium	**780 lb**	**780 kg**
1 meq Sodium	**460 lb**	**460 kg**

1 ppm = 1mg/kg = 2 pounds/acre = 2.24 kg/hectare

If TCEC is lower than 7, use value for 7. If it is over 39, use value for 39.

Date of Issue: 07/07/2012

Download full-size worksheet free at tinyurl.com/IntelligentGardener or soilanalyst.org.

Excess Cations Worksheet, page 2

	Deficit From other side of worksheet	**Application Limit** Per acre/year	**Quantity and Material to Add**	**S**	**Mg**	**Ca**
Sulfur **S**		110 lb Ag Sulfur	Reduce excess calcium with Ag Sulfur			
Phosphorus **P**		175 lb/ac elemental P	Use Soft Rock Phosphate if you are not willing to use Monoammonium phosphate.			
Calcium **Ca**		If below 68% calcium saturation use ag lime sufficient to 68%	If excess Mg, K or Na, use gypsum to satisfy any sulfur deficit. If excess Ca, do not use gypsum, use ag sulfur.			
Magnesium **Mg**		No more than 10% of target magnesium per year				
Potassium **K**		100 lb/ac elemental K				
Sodium **Na**						
Boron **B**		2 lb/ac elemental B				
Iron **Fe**						
Manganese **Mn**						
Copper **Cu**		No more than 7 lb elemental Cu				
Zinc **Zn**		No more than 14 lb elemental Zn				

	N	P	K	S	Ca	Mg
Fish Bone	4	8.8		.06	19.0	.03
Fish Meal	10	2		0.6	2.3	.03
Crab Shell	3	1.5	.025	.02	23.0	1.3
Blood Meal	13	0.5				
Feather Meal	12	0.0	0.35	0.4	0.6	
Bone Meal****	3	13.0		2.5	12.0	0.3
Oilseed Meal	6	1.5	1.0			
Copra Meal	4	1	0.7			
Kelp Meal	1	0.3	2.5	2	2	0.7
Ag Lime					32-39	2
Dolomite					22	13
Gypsum				17	20.5	
Oyster Shell					36	0.03
Magnesium Oxide						50
Montana Hard Rock Phos**		1.3			29	
Calphos		8.8			20	
Monoammonium Phosphate		23 (Plus 12% N as NH_3)				
K-Mag			18.2	22		11
Langbeinite			15.6	23		12
Greensand***		.05	6	1.3	1.5-3.0	2-4
Ag Sulphur				90		

Sea Salt	35% Sodium (Na)	
Borax	10% Boron (B)	
Iron Sulfate	18% S	30% Fe
Manganese sulfate	19% S	32% Mn
Copper Sulfate	12.5% S	25% Cu
Zinc Sulfate*	17% S	35% Zn
Potassium Sulfate	17% S	42% K
Magnesium Sulfate	13% S	10% Mg

* Zinc sulfate picks up moisture from the air; store in airtight container.

** Hard Rock Phosphate is 1.5% available P and contains around 27% insoluble phosphate.

*** Greensand contains 9% Fe, 50% Si and many trace elements. More than half its potassium content is insoluble.

**** Bonemeal contains 5.7% sodium.

Download full-size worksheet free at tinyurl.com/IntelligentGardener or soilanalyst.org.

Date of Issue: 07/07/2012

Calcareous Soil Worksheet

U.S. Measurements

Name ____________________

Plot or Field ____________________

Date of Test ____________________

Sample Depth 6 inches All numbers on this worksheet assume a six inch sample depth

pH ________ This worksheet is for soils that naturally hold free calcium (usually pH over 7.5) and those artifically created "Tiedjens" style (usually pH 7.1 or 7.2). Check that you actually have calcareous soil by doing a Fizz Test. Then get the proper soil test.

TCEC ________ Use the results from an elevated pH ammonium acetate extraction to determine TCEC. If necessary, calculate TCEC yourself using the formula on the bottom of this worksheet. Do not use levels discovered by a Mehlich 3 extraction or an ammonium acetate at pH 7.0 extraction to determine TCEC on calcareous or"over" limed soils.

Organic Matter % ________ From normal soil organic matter decomposition, assume approximate release of N = 15-25 lb N per 1% OM. Varies with temperature, moisture and soil air supply. N = 0.22 x NO_3

Element	Actual Level	Calculating Target Level	Target (Pounds per acre)	Deficit (Pounds per acre)
Sulfur **S**	ppm lb/ac	**S minimum = Mg (Target Level)**		
Phosphorus **P**	ppm lb/ac	**P = K (Target Level)** Calculate using actual P, not phosphate. P = 0.44 x P_2O_5		
Calcium **Ca**	ppm lb/ac	**TCEC x 400 x 0.85 = Target Level**		
Magnesium **Mg**	ppm lb/ac	**TCEC x 240 x 0.05 = Target Level**		
Potassium **K**	ppm lb/ac	**K is proportional to TCEC: see chart**		
Sodium **Na**	ppm lb/ac	**TCEC x 460 x 0.01 = Target Level** Be certain of good water quality before adding sodium		
Boron **B**	ppm lb/ac	**B = 2 lb/ac if CEC below 10** **= 4 lb/ac if CEC above 10**	Do not exceed 4 pounds	
Iron **Fe**	ppm lb/ac	**Fe = 100 lb/ac if CEC below 10** **= 150 lb/ac if CEC above 10**		
Manganese **Mn**	ppm lb/ac	**Mn = 55 lb/ac if CEC below 10** **= 100 lb/ac if CEC above 10**		
Copper **Cu**	ppm lb/ac	**Cu = ½ Zn (Target Level)**		
Zinc **Zn**	ppm lb/ac	**Zn = 1/10 P (Target Level)**		

Potassium Target Levels

TCEC	Pounds	TCEC	Pounds	TCEC	Pounds
		16	308	28	394
		17	316	29	397
		18	324	30	400
7	201	19	332	31	403
8	212	20	340	32	406
9	225	21	348	33	409
10	240	22	356	34	412
11	252	23	364	35	415
12	264	24	372	36	418
13	276	25	380	37	420
14	288	26	384	38	422
15	300	27	388	39	424

If TCEC is lower than 7, use value for 7. If it is over 39, use value for 39.

	One acre, six inches deep weighs	One hectare, 80 mm deep weighs
1 meq Calcium	**400 lb**	**400 kg**
1 meq Magnesium	**240 lb**	**240 kg**
1 meq Potassium	**780 lb**	**780 kg**
1 meq Sodium	**460 lb**	**460 kg**

1 ppm = 1mg/kg = 2 pounds/acre = 2.24 kg/hectare

Calculating TCEC:

$$\frac{\frac{\text{lb/ac calcium}}{400} + \frac{\text{lb/ac Mg}}{240} + \frac{\text{lb/ac K}}{780} + \frac{\text{lb/ac Na}}{460}}{(100 - \text{percent } H^+ - \text{other bases})} \times 100 = \text{TCEC}$$

In the case of calcareous soil, there is no H^+ and other bases usually are about 4%.

Download full-size worksheet free at tinyurl.com/IntelligentGardener or soilanalyst.org.

Calcareous Soil Worksheet, page 2

	Deficit From other side of worksheet	**Application Limit** Per acre/year	**Quantity and Material to add**	**S**	**Mg**	**Ca**
Sulfur **S**			If no other sulphates needed, use gypsum to reach minimum target level.			
Phosphorus **P**		175 lb/ac elemental P	Use Soft Rock Phosphate if you are not willing to use Monoammonium phosphate.			
Calcium **Ca**		Gypsum: 1 ton per acre	Use gypsum; it is okay to exceed minimum sulfur target.			
Magnesium **Mg**		No more than 20% of target magnesium per year	Use K-Mag or Langbeinite even if this puts K or S into excess.			
Potassium **K**		100 lb/ac elemental K	Use potassium sulphate. If this puts sulfur over the target level, go ahead anyway.			
Sodium **Na**						
Boron **B**		2 lb/ac elemental B				
Iron **Fe**		Foliar feeding only				
Manganese **Mn**		No more than 10 lb elemental Mn				
Copper **Cu**		No more than 5 lb elemental Cu				
Zinc **Zn**		No more than 10 lb elemental Zn				

	N	**P**	**K**	**S**	**Ca**	**Mg**
Fish Bone	4	8.8		.06	19.0	.03
Fish Meal	10	2		0.6	2.3	.03
Crab Shell	3	1.5	.025	.02	23.0	1.3
Blood Meal	13	0.5				
Feather Meal	12	0.0	0.35	0.4	0.6	
Bone Meal****	3	13.0		2.5	12	0.3
Oilseed Meal	6	1.5	1.0			
Copra Meal	4	1	0.7			
Kelp Meal	1	0.3	2.5	2	2	0.7
Ag Lime					32-39	2
Dolomite					22	13
Gypsum				17	20.5	
Oyster Shell					36	0.03
Magnesium Oxide						50
Montana Hard Rock Phos**		1.3			29	
Calphos		8.8			20	
Monoammonium Phosphate		23 (Plus 12% N as NH_3)				
K-Mag			18.2	22		11
Langbeinite			15.6	23		12
Greensand***		.05	6	1.3	1.5-3.0	2-4
Ag Sulphur				90		

Sea Salt	35%	Sodium (Na)
Borax	10%	Boron (B)
Iron Sulfate	18% S	30% Fe
Manganese sulfate	19% S	32% Mn
Copper Sulfate	12.5% S	25% Cu
Zinc Sulfate*	17% S	35% Zn
Potassium Sulfate	17% S	42% K
Magnesium Sulfate	13% S	10% Mg

* Zinc sulfate picks up moisture from the air; store in airtight container.

** Hard Rock Phosphate is 1.5% available P and contains around 27% insoluble phosphate.

*** Greensand contains 9% Fe, 50% Si and many trace elements. More than half its potassium content is insoluble.

**** Bonemeal contains 5.7% sodium.

Download full-size worksheet free at tinyurl.com/IntelligentGardener or soilanalyst.org.

Date of Issue: 07/07/2012

Index

Page numbers in *italics* indicate figures and tables.

About the Authors

Steve Solomon

Right now Steve is an active, alert 70 year old who grows vegetables and vegetable seeds on an entire quarter acre residential house block. His surplus veggies supply half a dozen local families with a weekly food box. He is mentoring two start-up vegetable garden seed businesses, makes presentations about gardening and health-related topics, moderates an active Yahoo e-mail chat group called "soilandhealth." His latest (ad)venture is serving as a neighbourhood soil analyst.

He spent much of his adult life living in rural Oregon, where he started Territorial Seed Company, a mail-order vegetables seed business, and sold it in 1986 to retire at age 44. Until her death in 1996 Steve was married to Dr. Isabelle Moser, a well known nature-curist who ran the Great Oaks School of Health. He co-authored a book with her titled *How and When To Be Your Own Doctor.* He is also the author of *Gardening When it Counts.*

Erica Reinheimer

Gardening has been Erica's life-long passion. Vivid early childhood memories include sharing flower and vegetable gardens with both sets of grandparents, and with her mom, an excellent garden designer. In her early 20's she plunged into organic vegetable gardening, moving to a remote 19 acre homestead with Alice, raising goats, chickens, and vegetables, and reading everything available on gardening. In her 30s and 40s she was gardening on a half acre in the Pacific Northwest, working in high tech engineering and R&D, and beginning to use seedmeal, soil testing and minerals in the garden — all with outstanding results. She started a home greenhouse business, and still runs a garden design business.

Rather suddenly, after 30 years of gardening, she was moved to share these years of accumulated experience. At about the same time, she began email conversations with Steve Solomon (and many others), sharing, modifying, and building a systematic method of soil analysis, mineralization, and horticulture. These conversations developed into parts of this book.

She now lives on an acre of rather thin (class 7 of 8) loamy sand on the central California coast, enjoying a really delicious garden because of the the techniques in this book, and helping others nourish their soils. And, she tries to remember everyday that as one shapes the garden, so does the garden shape you.

"It doesn't matter how the harvest will come out", says Masanobu Fukuoka."Just sow seeds and care tenderly for the plants and soil. You have joy. The ultimate goal of farming is not the growing of crops but the cultivation and perfection of human beings."

New Society Publishers

ENVIRONMENTAL BENEFITS STATEMENT

New Society Publishers has chosen to produce this book on recycled paper made with **100% post consumer waste,** processed chlorine free, and old growth free.

For every 5,000 books printed, New Society saves the following resources:[1]

35	Trees
3,126	Pounds of Solid Waste
3,439	Gallons of Water
4,486	Kilowatt Hours of Electricity
5,682	Pounds of Greenhouse Gases
24	Pounds of HAPs, VOCs, and AOX Combined
9	Cubic Yards of Landfill Space

[1]Environmental benefits are calculated based on research done by the Environmental Defense Fund and other members of the Paper Task Force who study the environmental impacts of the paper industry.